Phase Separation Coupled with Damage Processes

Christian Heinemann • Christiane Kraus

Phase Separation Coupled with Damage Processes

Analysis of Phase Field Models in Elastic Media

 Springer Spektrum

Christian Heinemann
Christiane Kraus

Weierstrass Institute for Applied
  Analysis and Stochastics
Berlin, Germany

Dissertation Humboldt-Universität zu Berlin, 2013

ISBN 978-3-658-05251-5   ISBN 978-3-658-05252-2 (eBook)
DOI 10.1007/978-3-658-05252-2

The Deutsche Nationalbibliothek lists this publication in the Deutsche Nationalbibliografie;
detailed bibliographic data are available in the Internet at http://dnb.d-nb.de.

Library of Congress Control Number: 2014933590

Springer Spektrum
© Springer Fachmedien Wiesbaden 2014

Printed on acid-free paper

Springer Spektrum is a brand of Springer DE.
Springer DE is part of Springer Science+Business Media.
www.springer-spektrum.de

Abstract

In applied analysis, mathematical frameworks for phase-field models to describe damage processes and phase separation in elastic media have been separately developed. This thesis is dedicated to a unifying model which couples both processes in a PDE system. The model has technological applications to solder joints where interactions of both phenomena have been observed and cannot be neglected for a realistic description. We develop suitable weak formulations for various types of this coupled system and prove existence of weak solutions.

Phase separation processes in this unifying approach are described by elastic Cahn-Hilliard or Allen-Cahn type equations. The corresponding chemical potential can either be of polynomial or logarithmic structure. The damage processes are modeled by a differential inclusion and are assumed to be rate-dependent and uni-directional, whereas the elastic deformations are specified by a quasi-static force balance law together with the small strain assumption. We first study cases in which the damage processes are incomplete, i.e., the maximally damaged regions still exhibit elastic properties. Global-in-time existence of weak solutions is proven for this non-degenerated case.

By performing a degenerate limit, we also investigate cases in which the damage evolution may lead to a complete disintegration of the material. The mobility tensor in the chemical diffusion equation now depends on the damage variable and is degenerating when the damage process is complete. The corresponding PDE problem is formulated via a time-dependent domain and translated into an SBV setting for a weak notion. We provide a maximal local-in-time existence result and global-in-time existence in some weaker sense.

The proofs of the existence results are based on several approximation schemes, a higher integrability property for the strain tensor, different variational techniques and regularization methods which will be established in this work.

Keywords: Cahn-Hilliard equations, Allen-Cahn equations, phase separation, damage processes, complete damage, elliptic-parabolic systems, coupled systems, nonlinear differential inclusions, weak solutions, existence results, rate-dependent systems

Zusammenfassung

Bislang wurden in der angewandten Analysis mathematische Theorien zu Phasenfeldmodellen, die Schädigungsprozesse und Phasenseparation in elastischen Medien beschreiben, getrennt entwickelt. Diese Doktorarbeit ist einem vereinheitlichenden Modell gewidmet, bei dem beide Prozesse in einem PDE-System gekoppelt sind. Das Modell hat technologische Anwendungen in der Beschreibung von Lotverbindungen, da hier Wechselwirkungen von beiden Phänomenen beobachtet wurden und diese für eine realistische Beschreibung nicht vernachlässigt werden können. Wir entwickeln eine geeignete schwache Formulierung für verschiedene Typen dieses gekoppelten Systems und beweisen die Existenz von schwachen Lösungen.

Phasenseparationsprozesse werden in diesem vereinheitlichenden Ansatz mit elastischen Cahn-Hilliard- oder Allen-Cahn-Gleichungen beschrieben. Das entsprechende chemische Potenzial kann entweder eine polynomielle oder eine logarithmische Struktur besitzen. Die Schädigungsprozesse werden mit einer Differenzialinklusion modelliert und als ratenunabhängig und unidirektional angenommen, wohingegen die elastischen Deformationen mit einem quasistatischen Kräftegleichgewicht in Kombination mit der Annahme kleiner Verzerrungen beschrieben werden. Zunächst studieren wir Fälle, bei denen die Schädigungen nicht vollständig verlaufen, d.h., die maximal geschädigten Bereiche weisen immer noch elastische Eigenschaften auf. Die zeitlich globale Existenz von schwachen Lösungen wird für diesen nicht-degenerierten Fall bewiesen.

Mittels eines Grenzüberganges untersuchen wir auch Fälle, bei denen die Evolution der Schädigung zu einer vollständigen Desintegration des Materials führen kann. Der Mobilitätstensor in der chemischen Diffusionsgleichung hängt nun von der Schädigungsvariablen ab und degeneriert, wenn der Schädigungsprozess vollständig ist. Das entsprechende PDE-Problem wird mit Hilfe eines zeitabhängigen Gebietes formuliert und für die schwache Formulierung in ein Problem mit SBV-Funktionen übersetzt. Wir zeigen Existenzresultate für zeitlich maximal lokale Lösungen und globale Lösungen in einem schwächeren Sinne.

Die Beweise der Existenzresultate basieren auf verschiedenen Approximationsmethoden, höheren Integrabilitätseigenschaften für den Verzerrungstensor, verschiedenen Variationstechniken und Regularisierungsmethoden, die in dieser Arbeit eingeführt werden.

Acknowledgements

Foremost, I would like to thank my advisor Dr. Christiane Kraus from the Weierstrass Institute in Berlin (WIAS) for introducing the topic to me and for providing me with the opportunity to work on my doctoral thesis as a member of her research group. I am grateful for her support, enthusiasm, the many inspiring discussions and for the productive collaboration. I also thank my advisor Prof. Dr. Jürgen Sprekels from WIAS for helpful discussions and for very useful advices.

I would like to acknowledge the Berlin Mathematical School (BMS) for financial support and for organizing meetings with fellow PhD students. In particular, I thank my BMS mentor Prof. Dr. Alexander Bockmayr from the Freie Universität Berlin for guidance and for his support during the preparation of this thesis.

I especially thank Wolfgang Giese from the Humboldt-Universität zu Berlin for proofreading certain parts of this work.

Finally, I express my sincere gratitude to my parents Elisabeth and Wolfgang Heinemann for their personal support.

Christian Heinemann

Contents

CHAPTER 1

Introduction

A better understanding of the mechanism and the interplay between phase separation and damage processes in elastically stressed solids is of big interest in material sciences. Various technological applications concerning the manufacturing and lifetime prediction of microelectronic devices are directly related to these phenomena. For example, solder joints in microelectronic packages connect the microchips to the circuit-boards and are, consequently, very critical components for the reliability engineering (see [LSC+04]). Solder materials usually consist of two or three component alloys whose aging process is influenced by temperature cycling. At high temperatures, solder alloys energetically favor one homogeneous phase consisting of a specific mixture of their chemical components. However, as soon as alloys are quenched sufficiently, phase separation or spinodal decomposition leads to fine-grained structures of different chemical compositions on a short time-scale. The long-term evolution is determined by a diffusion process which tends to minimize the bulk and the surface energies of the substances. J. W. Cahn and J. E. Hilliard developed a phenomenological model for the kinetics of phase separation in a thermodynamically consistent framework known as the Cahn-Hilliard equation [CH58]. They modeled the coarsening processes in solids by a fourth order parabolic equation which is mass conserving and can be expressed as an H^{-1}-gradient flow system. We refer to [Ell89] for analytical and numerical considerations as well as to [DM01] for experimental observations (see also Section 3.1 for further details and references).

The rate of coarsening and the morphology of the phases are influenced by their varying elastic properties whose contributions are given by different eigenstrains and elastic moduli as pointed out in [Gar00]. The coupled PDE system describing both phase separation and elasticity is sometimes referred to as the Cahn-Larché system. Beyond that, the different physical properties of the phases may lead to very high mechanical stresses preferably at interfacial regions (see [DM01, LSC+04]). These stress concentrations initiate the nucleation of microcracks and microvoids in solder materials whose propagation may eventually result in a complete failure of the whole device as investigated in [LSC+04, HCW91, USG07, GUE+07]. For mathematical literature about damage processes, we refer to Section 3.2, Section 3.3 as well as the references therein. Figure 1.1 and Figure 1.2 illustrate the chemical and mechanical processes involved in the interior of solder materials. They reveal mutual interferences of damage behavior driven by the growth of microfractures and microcavities with phase separation in alloys. In particular, we can observe that cracks prefer propagating along phase-boundaries between different chemical mixtures.

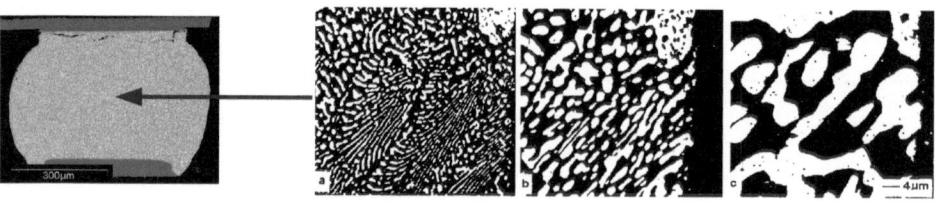

Figure 1.1: *Coarsening processes in binary alloys (see [HCW91]); Left: solder ball consisting of eutectic Sn–Pb; Right: (a) directly after solidification, (b) after 3 hours, and (c) after 300 hours.*

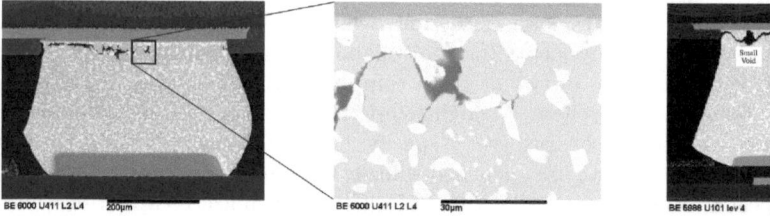
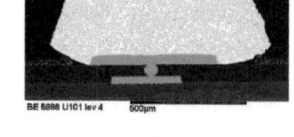

Figure 1.2: *Initiation and propagation of microcracks along phase boundaries (see [LSC+04]); Left, middle: crack-path with voids in a solder ball; Right: complete failure.*

This motivates the necessity for a unifying model and its rigorous analysis which is the topic of the present thesis. To our best knowledge, coupled PDE systems which describe damage phenomena, phase separation processes and elasticity have not been investigated in the mathematical literature so far. Even though both phase separation as well as damage processes have been subject to intensive research in the recent decades, they are treated separately. To each topic, rigorous mathematical tools have been developed to study existence, uniqueness, regularity and long-time behavior of solutions as well as numerical approximation schemes with finite element methods.

In the following, we will introduce the mathematical setting used throughout the subsequent chapters and the corresponding papers. Unless otherwise stated, we assume that the reference configuration of the elastic solid can be described by a bounded Lipschitz domain $\Omega \subseteq \mathbb{R}^n$ and its displacement field by a function u specifying the deformation

$x + u(x,t)$ of every material point $x \in \Omega$ at time $t \in (0, T)$. The function e signifies the strain tensor and σ the stress tensor with respect to the reference configuration. However, for the classical notions of solutions developed in Chapter 3, we assume Ω to be a bounded $\mathcal{C}^1$-domain. Then, the outer unit normal vector field on $\partial\Omega$ is denoted by ν.

Moreover, we assume a quasi-static force balance, i.e., the mechanical equilibrium is attained at all times, and the small strain assumption, i.e., the strain calculates as $e = \frac{1}{2}(\nabla u + (\nabla u)^t)$ (see, e.g., [HR99]). The behavior of the elastic solid might be influenced by many further physical processes such as, for instance, thermal effects [RR08], plasticity [HR99], adhesion [BBR12], damage processes [BSS05], magnetization [CVC12], phase separation [Ell89] or various phase change phenomena [Fré12], where each requires additional state quantities. In this thesis, as mentioned before, we will focus on a coupled system consisting of PDEs for phase separation and for damage processes in elastic solids and neglect other phenomena such as thermal effects. For both processes, smooth transitions between the different states are assumed and, therefore, phase-field formulations are used. More specifically (details can be found in Chapter 3):

- *Phase separation processes.* Phase separation and coarsening phenomena are described by Cahn-Hilliard or Allen-Cahn type equations. The order parameter for the chemical concentration $c : \Omega \times (0, T) \to \mathbb{R}^N$ is a vector function which specifies the ratio of each of the N constituents in the chemical mixture. In particular, the values of the components of c satisfy the constraints $c_1 + \ldots + c_N = 1$ and $c_i \geq 0$ for every $i = 1, \ldots, N$.

- *Damage processes.* The damage processes are modeled by a doubly nonlinear differential inclusion which incorporates a triggering mechanism: the damage only increases when the strain exceeds a critical value. The order parameter for the damage $z : \Omega \times (0, T) \to [0, 1]$ is a macroscopic scalar quantity which correlates to the volumetric fraction of undamaged material structures on a microscopic scale and is interpreted as follows: a value of $z(x) = 1$ indicates an undamaged and a value $z(x) = 0$ indicates a maximally damaged material point $x \in \Omega$. Intermediate values from $(0, 1)$ correspond to partially damaged material. Furthermore, the damage processes are considered as uni-directional, i.e., z is monotonically decreasing with respect to time.

The evolutionary PDEs for the coupled system with the unknowns c, u and z can be deduced from balance laws in continuum physics and from constitutive relations. Derivations for the uncoupled systems were proposed in [Gur96] for phase separation and in [FN96] for damage models. An approach for deriving the fully coupled PDE system is presented in Section 3.4.

The coupled system we are going to study is the following elliptic-parabolic system:

$$\partial_t c = \mathrm{div}(\mathbb{M}(z)\nabla\mu) \text{ with } \mu = -\mathrm{div}(\partial_{\nabla c}\psi) + \partial_c\psi \qquad \text{in } \Omega_T, \tag{1.1a}$$

$$0 = \mathrm{div}(\partial_e\psi(c,e,z)) \text{ with } e = (\nabla u + (\nabla u)^{\mathrm{t}})/2 \qquad \text{in } \Omega_T, \tag{1.1b}$$

$$0 \in -\mathrm{div}(\partial_{\nabla z}\psi) + \partial_z\psi + \partial\phi(\partial_t z) \qquad \text{in } \Omega_T \tag{1.1c}$$

with the unknowns c, u and z and the initial-boundary conditions

$$c(0) = c^0 \qquad\qquad \text{in } \Omega, \tag{1.2a}$$

$$z(0) = z^0 \qquad\qquad \text{in } \Omega, \tag{1.2b}$$

$$u = b \qquad\qquad \text{on } (\Gamma_{\mathrm{D}})_T, \tag{1.2c}$$

$$\partial_e\psi \cdot \nu = 0 \qquad\qquad \text{on } (\Gamma_{\mathrm{N}})_T, \tag{1.2d}$$

$$\partial_{\nabla z}\psi \cdot \nu = 0 \qquad\qquad \text{on } (\partial\Omega)_T, \tag{1.2e}$$

$$\partial_{\nabla c}\psi \cdot \nu = 0 \qquad\qquad \text{on } (\partial\Omega)_T, \tag{1.2f}$$

$$\mathbb{M}(z)\nabla\mu \cdot \nu = 0 \qquad\qquad \text{on } (\partial\Omega)_T, \tag{1.2g}$$

where $\psi = \psi(c, \nabla c, e, z, \nabla z)$ denotes the density of the free energy, $\phi = \phi(z_t)$ the density of the damage dissipation potential, μ the chemical potential and $\mathbb{M}$ the diffusion mobility. The Cahn-Hilliard equation is given by (1.1a), the balance of momentum equation is given by (1.1b) and the differential inclusion for the damage processes is given by (1.1c). The inclusion (1.1c) should be read in terms of generalized subdifferentials for $\partial_z\psi$ and $\partial\phi$, e.g. in our case $\psi = \psi^1 + \psi^2$ will split into a differentiable ψ^1 and a convex part ψ^2 (with respect to z) such that the generalized subdifferential reads as $\partial_z\psi^1 + \partial_z\psi^2$, where the latter ∂_z indicates the usual subdifferential operator for convex functions.

The main aim of this thesis is to provide existence of weak solutions for the above system for various cases resulting from different free energy densities ψ of the type

$$\psi(c, \nabla c, e, z, \nabla z) = \frac{1}{2}\Gamma\nabla c : \nabla c + \frac{1}{p}|\nabla z|^p + W^{\mathrm{el}}(c, e, z) + W^{\mathrm{ch}}(c) + f(z) + I_{[0,\infty)}(z). \tag{1.3}$$

Here, W^{el} signifies the elastic energy density, W^{ch} the chemical energy density, Γ the energy gradient tensor for the diffuse surface energy, f a damage dependent potential and $I_{[0,\infty)}$ the indicator function on $[0,\infty)$ to account for the constraint $z \geq 0$. The p-gradient term for z describes the local interactions for the damage processes and serves as a regularization term for the mathematical analysis (see [FN96, Fré12] for $p = 2$). The damage dissipation potential density ϕ is assumed to be of the form

$$\phi(z_t) = -\alpha z_t + \frac{\beta}{2}|z_t|^2 + I_{(-\infty,0]}(z_t), \qquad \alpha \geq 0, \beta > 0. \tag{1.4}$$

The indicator function $I_{(-\infty,0]}(z_t)$ accounts for the irreversibility condition. Therefore, $\partial_t z$ is forced to be non-positive which means that the damage process is uni-directional.

Since $\beta > 0$, the considered damage processes are referred to as *rate-dependent* (see [KRZ11]) and the well-developed framework of *rate-independent* systems does not apply here. For an introduction to rate-independent systems, we refer to [Mie05].

We can see from (1.3) that the coupling of all unknowns takes place in the elastic energy density W^{el}. Due to the effect of damage on the elastic response of the material, W^{el} is modeled by the following ansatz:

$$W^{el}(c, e, z) = \frac{1}{2}(g(z) + \delta)(e - e^{\star}(c)) : \mathbf{C}(c)(e - e^{\star}(c)), \qquad (1.5)$$

where $\mathbf{C}$ denotes the material stiffness, g the influence of the damage function (which is non-negative), $e^{\star}$ the eigenstrain (or stress free strain) and δ a non-negative constant. If the stiffness tensor is independent of the concentration, (1.5) is referred to as *homogeneous*, otherwise, (1.5) is called *inhomogeneous*.

The free energy density ψ in (1.3) unifies the energy densities used in elastic Cahn-Hilliard models (cf., e.g., [Gar00]) as well as in gradient-of-damage models with linear elasticity (cf., e.g., [KRZ11]) as pointed out in the following:

- *Cahn-Larché model:*

$$\psi^{CL}(c, e) = \frac{1}{2}\Gamma\nabla c : \nabla c + W^{ch}(c) + \frac{1}{2}(e - e^{\star}(c)) : \mathbf{C}(c)(e - e^{\star}(c)),$$

- *Gradient of damage model:*

$$\psi^{D}(e, z) = \frac{1}{p}|\nabla z|^p + \frac{1}{2}(g(z) + \delta)e : \mathbf{C}e + f(z) + I_{[0,\infty)}(z).$$

Moreover, we distinguish between the *incomplete damage* (or partial damage) and the *complete damage* case for the existence proofs. In partial damage models, δ is assumed to be a small positive constant, i.e., even the maximally damaged regions exhibit a small amount of elastic energy. The degenerated case $\delta = 0$ corresponds to complete damage models and is treated differently. For more details, we refer to Section 3.3.

The main existence results in this thesis are stated in Theorem 4.2.7, Theorem 5.2.6, Theorem 5.2.7, Theorem 7.2.5 and Theorem 7.2.4. Subsequently, we highlight some mathematical aspects in this work which were developed to make the coupled system (1.1)-(1.2) generally accessible for an analytical treatment and to overcome substantial mathematical difficulties.

- *Notions for weak solutions.* The damage differential inclusion (1.1a) involves two subdifferentials of indicator functions and nonlinearly coupled terms. Due to its intricate structure, the regularity of the solutions we will obtain from a-priori estimates are not sufficient to employ a classical differential inclusion formulation. Therefore, weaker notions of solutions are developed in Proposition 4.2.1, Proposition 5.2.1 and Theorem 6.2.3. Inspired by the concept of energetic solutions for rate-independent systems (see [Mie05]), the differential inclusion is expressed as a variational inequality together with a total energy inequality (see Definition 4.2.6). These notions of weak solutions form the basis for all existence results in this work.

- *Variational methods.* The existence proofs are based on a semi-implicit discretization scheme in time, where a recursive functional minimization is performed. These time-discrete solutions satisfy a discrete variational property for the damage evolution. We develop an approximation technique in Lemma 2.3.18 and a variational method in Lemma 2.3.19 to gain the correct variational inequality in the limit regime.

- *Regularization.* It turns out that the recursive minimization process yields a time-discrete energy estimate which is not sharp enough for the notion of weak solution in the continuous limit. We solved this problem by analyzing an appropriate regularization of the system in the first instance (see Definition 4.2.3 as well as Definition 5.2.2). In the time-discrete version of the regularized system, the chain-rule applied to the energy of the discrete solution curves yields the desired energy estimate together with an error term as shown in Lemma 4.3.10. However, in the regularized setting, the error term converges to 0 by passing from the time-discrete to the time-continuous regime. Finally, a further limit passage gives the correct energy inequality for the solutions.

Additionally, the following ideas have been developed for the complete damage case.

- *Time-dependent domain approach for complete damage.* During damage evolution processes, it might be possible that undamaged material fragments are surrounded by completely damaged material. In this case, due to analytical reasons, it is not possible to establish a displacement field on these parts. We circumvent this problem by formulating the whole evolution problem on a time-dependent domain. It encompasses all path-connected components of the not completely damaged area which are connected to the Dirichlet boundary (see Definition 3.4.3).

- *Lipschitz representation of non-smooth time-dependent domains.* Our considered complete damage system is regularized by incomplete damage systems in the first instance. In the passage to the limit, it is not possible to derive global a-priori estimates for the displacement fields due to missing uniform coercivity properties of the free energy. Lemma 2.4.6 provides a representation result which allows to control the displacements in some local sense by Korn's inequality (see also Lemma 6.3.13).

- *Degenerated damage-dependent chemical mobility.* The chemical potential μ in the Cahn-Hilliard equation (1.1a) depends on the strain (and, therefore, on the displacement variable) via $\partial_c W^{\mathrm{el}}$. Since the displacement field can only be established in the not completely damaged regions, we prevent phase separation processes on the completely damaged material by a degenerated mobility $\mathbb{M}$. More precisely, the diffusion mobility depends on the damage variable and vanishes when the damage is complete. With the help of the so-called *conical Poincaré inequality* (see [BK98]), the chemical potential can be established in the not completely damaged parts (see Lemma 7.3.5).

Structure of the thesis

The thesis is based on the papers [HK11, HK13, HK12a, HK12b].

Chapter 2 provides mathematical preliminaries which are used throughout this work. For instance, certain basic definitions in the theory of BV and SBV functions are summarized in Section 2.2. Then, in Section 2.3, we review some selected inequalities such as Korn's inequality and generalized versions of the Poincaré inequality as well as Aubin-Lions type embedding results. Additionally, we also present variational and approximation techniques which were first introduced in [HK11] (see Lemma 2.3.18 and Lemma 2.3.19). The study of the complete damage case requires deeper insights into time-dependent domains. A covering and representation result with Lipschitz domains as well as space-time local Sobolev spaces on these domains are treated in Section 2.4.

In Chapter 3, we will introduce phase-field models for phase separation as well as for damage processes in more detail. We will motivate the corresponding PDEs from physical balance laws and constitutive relations. Subsequently, we will turn our attention to the coupled PDE system (1.1) of damage and phase separation which has application to solder materials in microelectronic devices as mentioned before. Thermodynamic consistency of the PDE system is also shown.

The main part of this thesis – Chapter 4, Chapter 5, Chapter 6 and Chapter 7 – is devoted to the existence of weak solutions for systems of type (1.1) for various free energy densities ψ.

Chapter 4 is dedicated to the incomplete damage case where polynomial growth conditions for the chemical energy density function and homogeneous elastic energy densities with respect to the chemical concentration are assumed. We motivate an appropriate notion of weak solutions for the coupled PDE system. New variational, approximation and regularization techniques (cf. Chapter 2) are utilized to gain existence of weak solutions.

The existence results in Chapter 4 are extended in Chapter 5 to logarithmic chemical energy densities, inhomogeneous elastic energy density functions and quadratic damage gradients in ψ. To this end, the notion of solutions is slightly weakened. We employ a further regularization scheme and prove a higher integrability result for the strain tensor. Furthermore, Allen-Cahn systems coupled with damage processes are also considered.

Complete damage processes in purely mechanical systems with a quasi-static force balance are investigated in Chapter 6. A classical formulation of the PDE system with a time-dependent domain is motivated. We proceed with a weak formulation in an SBV setting, and using a characterization via variational inequalities, a weakened energy inequality and a jump condition. By performing a degenerate limit, maximal local-in-time existence of weak solutions is proven and global-in-time existence is shown in a weaker sense.

In Chapter 7, the results from the previous chapters are combined to analyze complete damage systems coupled with degenerating Cahn-Hilliard equations. Additional mathematical difficulties arise because the chemical mobility should depend on the damage variable and should vanish when the damage process is complete.

CHAPTER 2

Mathematical preliminaries

The following sections acquaint the reader with some mathematical background for the analytical part of this thesis. After the basic notations are given in Section 2.1, we will present some standard and nonstandard techniques for variational problems and since the presentation can, of course, not be comprehensive, we will refer to the monographs in functional analysis for more details at the corresponding parts.

A short introduction to measures and spaces of functions of bounded variations with values in a (possibly infinite dimensional) Banach space X is given in Section 2.2. These spaces will appear as trajectory spaces in weak notions of system (1.1) in Chapter 6 and Chapter 7. In particular, the space of SBV-functions defined on a time-interval and with values in X is introduced.

In Section 2.3, we firstly define the notion of Γ-convergence and summarize certain inequalities in Sobolev spaces and compact embedding results due to Aubin and Lions which are used in oder to obtain the proper convergence properties in the existence proofs. Beyond that, we introduce a new variational method in Subsection 2.3.3 and Subsection 2.3.4. More specifically, Lemma 2.3.19 gives in combination with an approximation scheme presented in Lemma 2.3.18 a new tool to deal with coupled variational inequalities arising from a weak formulation of the doubly nonlinear differential inclusion in (1.1c).

The analytical approach in Chapter 6 and Chapter 7 to complete damage systems requires certain spaces which do not seem to be well established in the mathematical literature so far. It turns out that the displacement field in a weak formulation of (1.1) exists only in a space-time local Sobolev space. This space is introduced in Section 2.4. Moreover, for the degenerate limit we employ covering and representation results for shrinking sets by families of Lipschitz domains which are proven in Section 2.4.

2.1   Notation

In this work, we fix a bounded Lipschitz domain $\Omega \subseteq \mathbb{R}^n$ of dimension n and a time constant $T > 0$. We assume that the material in the reference configuration is located in Ω. For the Dirichlet boundary Γ_D and the Neumann boundary Γ_N of $\partial\Omega$, we adopt the assumptions from [Ber11], i.e., Γ_D and Γ_N are non-empty and relatively open sets in $\partial\Omega$ with finitely many path-connected components such that $\Gamma_D \cap \Gamma_N = \emptyset$ and $\overline{\Gamma_D} \cup \overline{\Gamma_N} = \partial\Omega$. Note that $\mathcal{H}^{n-1}(\Gamma_D) > 0$). The following table provides an overview of some elementary notation used in this thesis.

- Measures and sets:

$\mathcal{L}^n, \mathcal{H}^n$	*n-dimensional Lebesgue and Hausdorff measure*
$\mathbb{R}_+, \mathbb{R}_\infty$	$[0, \infty),\ \mathbb{R} \cup \{+\infty\}$
$\mathbb{S}^n$	*n-dimensional unit sphere in* $\mathbb{R}^{n+1}$
Ω_T	$\Omega \times (0, T)$
$B_\varepsilon(A)$	*ε-neighborhood of* $A \subseteq \mathbb{R}^n$
$Q_\varepsilon(x_0)$	*open cube with center* $x_0 \in \mathbb{R}^n$ *and edge length* 2ε, *i.e.,* $\{x \in \mathbb{R}^n \mid \|x - x_0\|_\infty < \varepsilon\}$
$\overline{A}, \operatorname{int}(A), \partial A$	*closure, interior and boundary of* $A \subseteq \mathbb{R}^n$
$\{v = 0\}, \{v > 0\}$	*level and super-level set of v, i.e.,* $\{x \in \overline{\Omega} \mid v(x) = 0\}$ *and* $\{x \in \overline{\Omega} \mid v(x) > 0\}$ *for functions* $f \in L^1(\Omega)$ *defined up to a set of measure 0 and defined uniquely if* $v \in W^{1,p}(\Omega)$, $p > n$, *as* $W^{1,p}(\Omega) \hookrightarrow \mathcal{C}(\overline{\Omega})$
$\operatorname{supp}(v)$	*support of a function v, i.e.,* $\overline{\{x \mid v(x) \neq 0\}}$

- Spaces:

$\mathcal{L}(X)$	*space of linear and continuous functions from X to X*
$\mathcal{C}^k(\overline{\Omega}; \mathbb{R}^N)$	*space of k-times continuously differentiable functions on the open set $\Omega \subseteq \mathbb{R}^n$ where the k-th derivatives can be continuously extended to* $\overline{\Omega}$
$\mathcal{C}^{k,\alpha}(\overline{\Omega}; \mathbb{R}^N)$	*space of k-times continuously differentiable functions on the open set $\Omega \subseteq \mathbb{R}^n$ where the k-th derivatives are Hölder continuous with exponent α and can be continuously extended to* $\overline{\Omega}$

$\mathcal{C}_x^k(\overline{\Omega_T}; \mathbb{R}^N)$	*space of k-times continuously differentiable functions with respect to the spatial variable on the set $\Omega \times [0, T]$ where the k-th spatial derivatives can be continuously extended to $\overline{\Omega_T}$*
$W^{m,p}(\Omega; \mathbb{R}^N)$	*standard Sobolev space of m-times weakly differentiable functions with weak derivatives in $L^p(\Omega; \mathbb{R}^N)$*
$W_+^{1,r}(\Omega), W_-^{1,r}(\Omega)$	*space of non-negative and non-positive Sobolev functions, i.e., $\{\zeta \in W^{1,r}(\Omega) \,\vert\, \zeta \geq 0 \text{ a.e. in } \Omega\}$ and $\{\zeta \in W^{1,r}(\Omega) \,\vert\, \zeta \leq 0 \text{ a.e. in } \Omega\}$*
$W_{\Gamma_D}^{1,r}(\Omega; \mathbb{R}^N)$	*space of Sobolev functions vanishing on the Dirichlet boundary Γ_D: $\{\zeta \in W^{1,r}(\Omega; \mathbb{R}^N) \,\vert\, \zeta = 0 \text{ on } \Gamma_D \text{ in the sense of traces}\}$*
X^*	*dual space of the Banach space X*

- Functions and operators:

$\mathbb{1}_A, I_A$	*characteristic function and indicator function $X \to \mathbb{R}_\infty$ with respect to a subset $A \subseteq X$*
$A : B$	*Euclidean matrix product of $A \in \mathbb{R}^{n \times n}$ and $B \in \mathbb{R}^{n \times n}$*
$\langle \cdot, \cdot \rangle_{X^\star \times X}$	*dual pairing of $X^\star$ and X, abbr. $\langle \cdot, \cdot \rangle$*
$[f]^+$	*non-negative part of f, i.e., $\max\{0, f\}$*
f^+, f^-	*one-sided limits of $f : I \to X$, $I \subseteq \mathbb{R}$ interval, i.e., $f^\pm(a) = \lim_{x \to a\pm} f(x)$*
Δ_p	*p-Laplacian $\Delta_p := \text{div}(\vert\nabla z\vert^{p-2} \nabla z)$*
$\fint_A f(x)\,\mathrm{d}x$	*mean value of f in $A \subseteq \mathbb{R}^n$, i.e., $\frac{1}{\mathcal{L}^n(A)} \int_A f(x)\,\mathrm{d}x$*
∂J	*subdifferential of a convex function $J : X \to \mathbb{R}_\infty$ i.e., $\partial J(x) = \{x^\star \,\vert\, J(x) + \langle x^\star, y - x \rangle \leq J(y) \text{ for all } y\}$*
$\mathrm{d}E$	*Gâteaux differential of a functional $E : X \to Y$*
$p^\star$	*Sobolev critical exponent $\frac{np}{n-p}$ for $n > p$*
$\text{diam}(Q)$	*diameter of a subset $Q \subseteq \mathbb{R}^n$*

- Binary relations:

$A \subset\subset B$	*if $\overline{A} \subseteq B$*

We adopt the convention that for two given functions $\zeta, \xi \in L^1(0, T; W^{1,p}(\Omega))$ with $p > n$ the inclusion $\{\zeta = 0\} \supseteq \{\xi = 0\}$ is an abbreviation for $\{\zeta(t) = 0\} \supseteq \{\xi(t) = 0\}$ for a.e. $t \in (0, T)$. Here, $\zeta(t), \xi(t) \in \mathcal{C}(\overline{\Omega})$ due to the embedding $W^{1,p}(\Omega) \hookrightarrow \mathcal{C}(\overline{\Omega})$.

2.2 Vector measures and vector-valued functions of bounded variations

In the beginning, we will review some basic definitions from the theory of vector measures and standard Bochner spaces. For further readings on this topic, we would like to refer to [Din02, CV02].

Let $(X, \|\cdot\|)$ be a Banach space and (S, Σ, μ) be a measure space consisting of a set S, a σ-algebra over S and a positive measure $\mu : \Sigma \to [0, +\infty]$. The measure space (S, Σ, μ) is called finite if $\mu(S) < \infty$. Furthermore, let $m : \Sigma \to X$ be a Banach space valued measure. We assume that all measures are σ-additive (finitely additive measures also studied in the literature; see [Din02, Chapter 1, §2.A]), i.e.,

$$\mu\left(\bigcup_{i \in \mathbb{N}} A_i\right) = \sum_{i \in \mathbb{N}} \mu(A_i) \quad \text{for all pairwise disjoint sets } A_i \in \Sigma, \, i \in \mathbb{N}.$$

As usual, $L^p(S, \mu; X)$ denotes the p-Bochner μ-integrable functions with values in X (μ-essentially bounded for $p = \infty$, respectively), cf. [Zei90, Chapter 23.2]. In the following, we restrict ourselves to the case where S is a finite interval $I \subseteq \mathbb{R}$. We write $L^p(I; X)$ for $L^p(I, \mathcal{L}^1; X)$. The subspace $H^q(I; X) \subseteq L^2(I; X)$, $q \in \mathbb{N}$, indicates the $L^2(I; X)$-functions f which are q-times weakly differentiable with weak derivatives $\partial_t^s f$ in $L^2(I; X)$, $s = 1, \ldots, q$, i.e.,

$$\int_I f(t) \partial_t^s \xi(t) \, dt = (-1)^s \int_I (\partial_t^s f(t)) \xi(t) \, dt \quad \text{for all } \xi \in C_0^\infty([0, T]).$$

The norm is given by $\|f\|_{H^q(I;X)} := \sum_{s=0}^q \|\partial_t^s f\|_{L^2(I;X)}$ (see [Zei90, Chapter 23.5]).

In the most literature, functions of bounded variations are usually considered with values in $\mathbb{R}$ or in a finite dimensional vector space [AFP00]. We will give the definition of BV and SBV-functions defined on a time-interval and with values in a Banach space X. X-valued BV-functions are, for instance, also investigated in the monograph [Din66].

Definition 2.2.1 (BV-space of Banach space-valued functions)
The subspace $BV(I; X) \subseteq L^1(I; X)$ consists of functions $f \in L^1(I; X)$ with the norm

$$\text{ess var}_I(f) := \inf \left\{ \text{var}_I(g) \, | \, g = f \, \mathcal{L}^1\text{-a.e. in } I \right\} < +\infty,$$

and

$$\text{var}_I(f) := \sup \left\{ \sum_{i=1}^{k-1} \|f(t_{i+1}) - f(t_i)\|_X \, \Big| \, t_1 < t_2 < \ldots < t_k \text{ with } t_1, t_2, \ldots t_k \in I \text{ for } k \geq 2 \right\}.$$

Since the weak derivatives of BV-functions are measures, we turn our attention to X-valued measures. To proceed, we define the variation of the vector-valued measure m as (cf. [Din02, Chapter 1, §2.A])

$$|m| := \text{var}(m) := \sup \left\{ \sum_{i \in I} \|m(A_i)\|_X \, \Big| \, \{A_i\} \text{ is a finite family of disjoint sets } A_i \in \Sigma \right\}.$$

We say that m has finite variation if $|m| < \infty$. Furthermore, m is said to be absolutely continuous with respect to μ, abbr. $m \ll \mu$, if for all $A \in \Sigma$ with $\mu(A) = 0$ we have $m(A) = 0$ (cf. [Din02, Chapter 1, §3.C]).

Now, let $f \in L^1(I, \mu; X)$ and $B \subseteq I$ be measurable. we define the measures $f\mu$ and $\mu\lfloor B$ as follows:

$$(f\mu)(A) := \int_A f(t)\,\mathrm{d}\mu(t) \text{ for measurable } A \subseteq I,$$

$$(\mu\lfloor B)(A) := \mu(A \cap B) \text{ for measurable } A \subseteq I.$$

For finite dimensional spaces X, there exists always a decomposition $m = g\mu$ for a function $g \in L^1(S, \mu; X)$, i.e., $m(A) = \int_A g\,\mathrm{d}\mu$ for all $A \in \Sigma$, when $m \ll \mu$ holds (see [AFP00, Theorem 1.28]). In this case, we call g the *Radon-Nikodým derivative*. This is, in general, not true when X is an infinite dimensional Banach space and, therefore, motivates the subsequent definition.

Definition 2.2.2 (Radon-Nikodým property, cf. [Din02, Chapter 1, §2.G])
A Banach space X has the Radon-Nikodým property if for every finite measure space (S, Σ, μ) and every measure $m : \Sigma \to X$ with finite variation $|m|$ such that $m \ll \mu$ there exists a function $g \in L^1(S, \mu; X)$ such that $m = g\mu$, i.e., $m(A) = \int_A g\,\mathrm{d}\mu$ for all $A \in \Sigma$.

Remark 2.2.3 *Reflexive spaces and separable duals of Banach spaces are examples for Banach spaces which possess the Radon-Nikodým property, see [Din02, Chapter 1, §2.G].*

To every $f \in BV(I; X)$, we can choose a representant (also denoted by f) with $\mathrm{var}_I(f) < +\infty$. Then the values $f(t^{\pm}) := \lim_{s \to t^{\pm}} f(s)$ exist for all $t \in \bar{I}$ (and are independent of the representant) by adopting the convention $f((\inf I)^-) := f((\inf I)^+)$ and $f((\sup I)^+) := f((\sup I)^-)$. The functions $f^+(t) := f(t^+)$ and $f^-(t) := f(t^-)$ are thus uniquely defined for every $t \in \bar{I}$ and do not coincide for at most countably many points, i.e., in the jump discontinuity set J_f. Furthermore, a regular measure $\mathrm{d}f$ with finite variation, i.e., $|\mathrm{d}f|(I) < \infty$, and with values in X (also called *differential measure*) can be assigned such that $\mathrm{d}f((a, b]) = f^+(b) - f^+(a)$ for all $a, b \in \bar{I}$ with $a \leq b$, cf. [Din66, Chapter III, §17.2, Theorem 1].

If X exhibits the Radon-Nikodým property the differential measure decomposes into $\mathrm{d}f = f'_\mu \mu$ for a positive Radon measure μ and a function $f'_\mu \in L^1(I, \mu; X)$ (see also [MV87]). We call f a special function of bounded variations if $\mathrm{d}f$ even decomposes into an *absolutely continuous part* and a *jump part*. More precisely, we define the following subspace (cf. [AFP00, Chapter 4.1]).

Definition 2.2.4 (SBV^p-space of vector-valued functions)
The subspace $SBV(I; X) \subseteq BV(I; X)$ of special functions of bounded variations is defined as the space of functions $f \in BV(I; X)$ where the decomposition

$$\mathrm{d}f = f'\mathcal{L}^1 + (f^+ - f^-)\mathcal{H}^0\lfloor J_f \tag{2.1}$$

for an $f' \in L^1(I; X)$ exists. Here, $\mathcal{H}^0$ denotes the 0-dimensional Hausdorff measure, i.e., (2.1) reads as

$$\mathrm{d}f(A) = \int_A f'(t)\,\mathrm{d}t + \sum_{t \in J_f \cap A} (f^+(t) - f^-(t)).$$

The function f' is called the absolutely continuous part of the differential measure and we also write $\partial_t^a f$. If, additionally, $\partial_t^a f \in L^p(I; X)$, $p \geq 1$, we write $f \in SBV^p(I; X)$.

Theorem 2.2.5 (BV-chain rule [MV87, Theorem 3]) *Let $I \subseteq \mathbb{R}$ be an interval, X be a Banach space with the Radon-Nikodým property, $f \in BV(I; X)$ with $\mathrm{d}f = f'_\mu \mu$ for a non-negative Radon measure μ on I and $f'_\mu \in L^1(I, \mu; X)$. Moreover, let $E : X \to \mathbb{R}$ be continuously Fréchet-differentiable. Then $E \circ f \in BV(I; \mathbb{R})$ and $\mathrm{d}(E \circ f)$ admits as density relative to μ the function $t \mapsto \langle \theta(t), f'_\mu(t) \rangle$, where $\theta : I \to X^\star$ is defined as*

$$\theta(t) := \int_0^1 \mathrm{d}E((1 - r)f(t^-) + rf(t^+))\,\mathrm{d}r.$$

Corollary 2.2.6 *Suppose $f \in SBV(0, T; X)$ and $E : X \to \mathbb{R}$ is continuously Fréchet-differentiable. Then $E \circ f \in SBV(0, T)$ and for all $0 \leq a \leq b \leq T$:*

$$\mathrm{d}(E \circ f)((a, b]) = \int_a^b \langle \mathrm{d}E(f(s)), f'(s) \rangle\,\mathrm{d}s + \sum_{s \in J_f \cap (a, b]} \left(E(f(s^+)) - E(f(s^-)) \right).$$

Proof. We apply Theorem 2.2.5. By assumption, we obtain the decomposition $\mathrm{d}f = f'_\mu \mu$ with $\mu = \mathcal{L}^1 + \mathcal{H}^0 \lfloor J_f$ and

$$f'_\mu(t) = \begin{cases} f'(t) & \text{if } t \in [0, T] \setminus J_f, \\ f(t^+) - f(t^-) & \text{if } t \in J_f. \end{cases}$$

Applying Theorem 2.2.5 yields

$$\mathrm{d}(E \circ f)((a, b]) = \int_{(a, b]} \langle \theta(s), f'_\mu(s) \rangle\,\mathrm{d}\mu(s)$$

$$= \int_{(a, b]} \langle \theta(s), f'(s) \rangle\,\mathrm{d}\mathcal{L}^1(s) + \sum_{t \in J_f \cap (a, b]} \langle \theta(s), f(s^+) - f(s^-) \rangle$$

Since $f(s^+) = f(s^-) = f(s)$ for $\mathcal{L}^1$−a.e. $s \in (a, b)$, the first term on the right hand side becomes

$$\int_{(a, b]} \langle \theta(s), f'(s) \rangle\,\mathrm{d}\mathcal{L}^1(s) = \int_{(a, b]} \left\langle \int_0^1 \mathrm{d}E((1 - r)f(s^-) + rf(s^+))\,\mathrm{d}r, f'(s) \right\rangle \mathrm{d}\mathcal{L}^1(s)$$

$$= \int_{(a, b]} \langle \mathrm{d}E(f(s), f'(s) \rangle\,\mathrm{d}s,$$

where, as usual, $ds := d\mathcal{L}^1(s)$. Furthermore, by the classical chain rule,

$$\sum_{s \in J_f \cap (a,b]} \langle \theta(s), f(s^+) - f(s^-) \rangle$$

$$= \sum_{s \in J_f \cap (a,b]} \left\langle \int_0^1 dE((1-r)f(s^-) + rf(s^+)) \, dr, f(s^+) - f(s^-) \right\rangle$$

$$= \sum_{s \in J_f \cap (a,b]} \int_0^1 \left\langle dE((1-r)f(s^-) + rf(s^+)), f(s^+) - f(s^-) \right\rangle dr$$

$$= \sum_{s \in J_f \cap (a,b]} \int_0^1 \frac{d}{dr} E((1-r)f(s^-) + rf(s^+)) \, dr$$

$$= \sum_{s \in J_f \cap (a,b]} \left(E(f(s^+)) - E(f(s^-)) \right).$$

$\square$

2.3 Variational methods

2.3.1 Γ-convergence

In the following, we give the definition of Γ-convergence and some basic properties taken from [Bra06]. For further details, we also refer to the monograph [Mas93, Bra02]. The complete damage approach in Chapter 6 as well as in Chapter 7 uses Γ-convergence methods to gain the energy estimate in the notion of weak solutions.

The definition of Γ-convergence can be given in the case of topological spaces. To this end, we fix a topological space $(X, \mathcal{T})$ and consider a sequence of functionals $\{f_\varepsilon\}$ with $f_\varepsilon : X \to [-\infty, +\infty]$, $\varepsilon \in (0,1)$, as well as $f : X \to [-\infty, +\infty]$. We say that f_ε Γ-converges to the functional $f : X \to [-\infty, +\infty]$ as $\varepsilon \to 0^+$ if for all $x \in X$

$$f(x) = \sup_{U \in \mathcal{N}(x)} \liminf_{\varepsilon \to 0} \inf_{y \in U} f_\varepsilon(y) = \sup_{U \in \mathcal{N}(x)} \limsup_{\varepsilon \to 0} \inf_{y \in U} f_\varepsilon(y)$$

is fulfilled, where $\mathcal{N}(x)$ denotes the set of all neighborhoods $U \in \mathcal{T}$ of x in X. In this case, we write $f_\varepsilon \xrightarrow{\Gamma} f$.

For further studies, we switch to metric spaces (X, d). We obtain a more convenient definition of Γ-convergence in this case. Note that bounded subsets of Banach spaces with separable duals are metrizable in the weak topology (see [AB07, Theorem 6.31]).

Theorem 2.3.1 (Γ-convergence, cf. [Bra06, Theorem 2.1])
The following properties are equivalent:

(i) $f_\varepsilon \xrightarrow{\Gamma} f$

(ii) $f = \Gamma - \liminf_{\varepsilon \to 0} f_\varepsilon = \Gamma - \limsup_{\varepsilon \to 0} f_\varepsilon$ *with*

$$\Gamma - \liminf_{\varepsilon \to 0} f_\varepsilon(x) := \inf\{\liminf_{\varepsilon \to 0} f_\varepsilon(x_\varepsilon) \mid x_\varepsilon \to x\}$$
$$\Gamma - \limsup_{\varepsilon \to 0} f_\varepsilon(x) := \inf\{\limsup_{\varepsilon \to 0} f_\varepsilon(x_\varepsilon) \mid x_\varepsilon \to x\}.$$

(iii) (a) liminf estimate. For every sequence $x_\varepsilon \to x$ in X, it holds

$$f(x) \leq \liminf_{\varepsilon \to 0} f_\varepsilon(x_\varepsilon).$$

(b) limsup estimate. There exists a sequence (so-called recovery sequence) $x_\varepsilon \to x$ in X such that

$$f(x) \geq \limsup_{\varepsilon \to 0} f_\varepsilon(x_\varepsilon).$$

At this point, we would like to mention the fundamental theorem of Γ-convergence which emphasizes its importance in variational methods (cf. [Bra06, Theorem 2.10]).

Theorem 2.3.2 (Fundamental theorem of Γ-convergence) *Assume $f_\varepsilon \xrightarrow{\Gamma} f$. Furthermore, let f_ε satisfy the following condition (equi-coercivity)*

$$\forall t \in \mathbb{R}, \; \exists K \subseteq X \text{ compact}, \; \forall \varepsilon \in (0,1) : \{f_\varepsilon \leq t\} \subseteq K.$$

Then f has a minimum in X and $\min_{x \in X} f(x) = \lim_{\varepsilon \to 0} \inf_{x \in X} f_\varepsilon(x)$.

In order to draw some further conclusions, we introduce the lower semi-continuous envelope $\overline{f} : X \to [-\infty, +\infty]$ of a functional f as

$$\overline{f}(x) := \liminf_{y \to x} f(y).$$

Remark 2.3.3 *We have the following properties (cf. [Bra06, Proposition 2.4 and Remark 2.12]).*

(i) If $f_\varepsilon \xrightarrow{\Gamma} f$ then f is lower semi-continuous.

(ii) If a sequence $\{f_k\}_{k \in \mathbb{N}}$ is monotonically decreasing, i.e., $f_{k+1} \leq f_k$ for all $k \in \mathbb{N}$, then the Γ-limit exists and is given by $\Gamma - \lim_{k \to \infty} f_k = \overline{\inf_{k \in \mathbb{N}} f_k}$.

2.3.2 Embedding theorems and inequalities

In the following, we give a short collection of some inequalities and compactness results which are extensively used in the successive chapters.

Theorem 2.3.4 (Sobolev embedding theorem)

(i) *Into Sobolev spaces [Alt99, Chapter 8.9]. Let $m_1, m_2, p_1, p_2 \in \mathbb{R}$ be constants with $m_1 > m_2 \geq 0$, $1 \leq p_1, p_2 < \infty$, $m_1 - \frac{n}{p_1} > m_2 - \frac{n}{p_2}$. Then, there exists the compact embedding*

$$\mathrm{Id} : W^{m_1, p_1}(\Omega) \hookrightarrow W^{m_2, p_2}(\Omega).$$

In the limiting case $m_1 - \frac{n}{p_1} = m_2 - \frac{n}{p_2}$ or $m_1 = m_2$, the embedding also exists and is continuous.

(ii) *Into Hölder spaces [Alt99, Chapter 8.5]. Let $m, p, k, \alpha \in \mathbb{R}$ be constants with $m \geq 1$, $1 \leq p < \infty$, $k \geq 0$, $0 < \alpha < 1$, $m - \frac{n}{p} > k + \alpha$. Then, there exists the compact embedding*

$$\mathrm{Id} : W^{m, p}(\Omega) \hookrightarrow C^{k, \alpha}(\overline{\Omega}).$$

In the limiting case $m - \frac{n}{p} = k + \alpha$, the embedding also exists and is continuous.

Theorem 2.3.5 (Poincaré's inequalities)
Let $p \geq 1$. There exists a $C > 0$ such that:

(i) *For functions with vanishing mean value [Zie89, Theorem 4.4.2].*

$$\int_\Omega \left| u - \fint_\Omega u \, \mathrm{d}x \right|^p \mathrm{d}x \leq C \int_\Omega |\nabla u|^p \, \mathrm{d}x \qquad \text{for all } u \in W^{1,p}(\Omega).$$

(ii) *For functions vanishing on a Dirichlet boundary of positive measure [Dob07, Theorem 6.22].*

$$\int_\Omega |u|^p \, \mathrm{d}x \leq C \int_\Omega |\nabla u|^p \, \mathrm{d}x \qquad \text{for all } u \in W^{1,p}_{\Gamma_\mathrm{D}}(\Omega).$$

Combining the Sobolev embedding theorem and the Poincaré's inequalities, we obtain the so-called Sobolev-Poincaré's inequalities. For our analysis, we will use the following versions.

Theorem 2.3.6 (Sobolev-Poincaré's inequality)
Let $1 \leq p < n$. There exists a constant $C > 0$ such that

(i) *for all rectangles $Q \subseteq \mathbb{R}^n$ and all $u \in W^{1,p}(Q)$:*

$$\left(\fint_Q \left| u - \fint_Q u \right|^{p^\star} \right)^{\frac{1}{p^\star}} \leq C \left(\fint_Q |\nabla u|^p \right)^{\frac{1}{p}} (\mathrm{diam}\, Q),$$

(ii) *for all rectangles $Q = \prod_{i=1}^{n}(a_i, b_i) \subseteq \mathbb{R}^n$ and all $u \in W^{1,p}(Q)$ with $u = 0$ on $\{(x_1, \ldots, x_{n-1}, a_n) \,|\, a_i \leq x_i \leq b_i,\ i = 1, \ldots, n-1\} \subseteq \partial Q$ (in the sense of traces):*

$$\left(\fint_Q |u|^{p^\star} \right)^{\frac{1}{p^\star}} \leq C \left(\fint_Q |\nabla u|^p \right)^{\frac{1}{p}} (\mathrm{diam}\, Q).$$

Remark 2.3.7 *Theorem 2.3.6 can be obtained by establishing the corresponding inequalities on the unit cube $(0,1)^n$ (by applying Theorem 2.3.4 and Theorem 2.3.5) and then using a scaling argument. It should be remarked that the case $1 < p < n$ has been considered by Sobolev [Sob38] while Nirenberg [Nir59] has studied the case $p = 1$.*

Theorem 2.3.8 (Korn's first inequality, cf. [Nef02])
There exists a $C > 0$ such that for all $u \in W^{1,p}_{\Gamma_D}(\Omega)$:

$$\int_\Omega (|u|^2 + |\nabla u|^2)\,\mathrm{d}x \leq C \int_\Omega |\epsilon(u)|^2\,\mathrm{d}x,$$

where $\epsilon(u)$ is defined as $\frac{1}{2}(\nabla u + (\nabla u)^{\mathrm{t}})$.

Theorem 2.3.9 (Aubin-Lions type embeddings, cf. [Sim86])
Let $X \subseteq B \subseteq Y$ be Banach spaces where X compactly embeds into B.

(i) For L^p-spaces. Let a sequence $\{f_k\}$ be bounded in $L^p(0,T;X)$ with $1 \leq p \leq \infty$ and the derivatives $\{\partial_t f_k\}$ be bounded in $L^p(0,T;Y)$. Then $\{f_k\}$ is relatively compact in $L^p(0,T;B)$.

(ii) For $\mathcal{C}$-spaces. Let a sequence $\{f_k\}$ be bounded in $L^\infty(0,T;X)$ and the derivatives $\{\partial_t f_k\}$ be bounded in $L^2(0,T;Y)$. Then $\{f_k\}$ is relatively compact in $\mathcal{C}([0,T];B)$.

The following theorem is an adaption from [Gia83, Chapter V.1, Proposition 1.1].

Theorem 2.3.10 (Reverse Hölder inequality, cf. [Gar00, Proposition 8.1])
Let $Q \subseteq \mathbb{R}^n$ be a cube, $g \in L^q_{\mathrm{loc}}(Q)$ for some $q > 1$ and $g \geq 0$. Suppose that there exist a constant $b > 0$ and a function $f \in L^r_{\mathrm{loc}}(Q)$ with $r > q$ and $f \geq 0$ such that

$$\fint_{Q_R(x_0)} g^q\,\mathrm{d}x \leq b \left(\fint_{Q_{2R}(x_0)} g\,\mathrm{d}x \right)^q + \fint_{Q_{2R}(x_0)} f^q\,\mathrm{d}x$$

for each $x_0 \in Q$ and all $R > 0$ with $2R < \mathrm{dist}(x_0, \partial Q)$. Then $g \in L^s_{\mathrm{loc}}(Q)$ for $s \in [q, q+\varepsilon)$ with some $\varepsilon > 0$ and

$$\left(\fint_{Q_R(x_0)} g^s\,\mathrm{d}x \right)^{\frac{1}{s}} \leq c \left(\left(\fint_{Q_{2R}(x_0)} g^q\,\mathrm{d}x \right)^{\frac{1}{q}} + \left(\fint_{Q_{2R}(x_0)} f^s\,\mathrm{d}x \right)^{\frac{1}{s}} \right)$$

for all $x_0 \in Q$ and $R > 0$ such that $Q_{2R}(x_0) \subseteq Q$. The positive constants $c, \varepsilon > 0$ depend on b, q, n and r.

Theorem 2.3.11 (Conical Poincaré inequality, cf. [BK98, Corollary 2])
Suppose that $\Omega \subseteq \mathbb{R}^n$ is a bounded and star-shaped domain, $r \geq 0$ and $1 \leq p < \infty$. Then, there exists a constant $C = C(\Omega, p, r) > 0$ such that

$$\int_\Omega |w(x) - w_{\Omega, \delta^t}|^p \delta^r(x)\,\mathrm{d}x \leq C \int_\Omega |\nabla w(x)|^p \delta^r(x)\,\mathrm{d}x$$

for all $w \in C^1(\Omega)$ where the δ^r-weight w_{Ω,δ^r} is given by

$$w_{\Omega,\delta^r} := \int_\Omega w(x)\delta^r(x)\,\mathrm{d}x, \quad \delta(x) := \mathrm{dist}(x, \partial\Omega).$$

Remark 2.3.12 *By a density argument, the statement is, of course, also true for all $w \in W^{1,p}(\Omega)$ which will be used in this paper.*

The inclusion $L^\infty(0,T;W^{1,p}(\Omega)) \cap H^1(0,T;L^2(\Omega)) \subseteq \mathcal{C}(\overline{\Omega_T})$ for $p > n$ follows from Theorem 2.3.9 (ii). It can also be shown with the following generalized version of Poincaré's inequality.

Theorem 2.3.13 (Generalized Poincaré inequality, cf. [Alt99, Theorem 6.15])

Let $M \subseteq W^{1,p}(\Omega;\mathbb{R}^m)$ be non-empty, convex and closed with $1 < p < \infty$. Furthermore, M satisfies the property

$$u \in M, \ \alpha \geq 0 \implies \alpha u \in M.$$

Then the following statements are equivalent:

(i) There exists a $u_0 \in M$ and a constant $C_0 > 0$ such that for all $\xi \in \mathbb{R}^m$

$$u_0 + \xi \in M \implies |\xi| \leq C_0.$$

(ii) There exists a constant $C > 0$ such that for all $u \in M$

$$\|u\|_{L^p(\Omega;\mathbb{R}^m)} \leq C\|\nabla u\|_{L^p(\Omega;\mathbb{R}^{m \times n})}.$$

Proposition 2.3.14 *Let $p > n$ be a constant. Then*

$$L^\infty(0,T;W^{1,p}(\Omega)) \cap H^1(0,T;L^2(\Omega)) \subseteq \mathcal{C}(\overline{\Omega_T}).$$

Proof. Let $z \in L^\infty(0,T;W^{1,p}(\Omega)) \cap H^1(0,T;L^2(\Omega))$. We can choose a representative such that $z \in \mathcal{C}([0,T];L^2(\Omega))$ and $z(t) \in W^{1,p}(\Omega)$ for all $t \in [0,T]$. By employing the embedding $W^{1,p}(\Omega) \subseteq \mathcal{C}(\overline{\Omega})$ (note that $p > n$), we obtain a representant $z : \overline{\Omega_T} \to \mathbb{R}$ such that

$$z \in \mathcal{C}([0,T];L^2(\Omega)) \text{ and } z(t) \in \mathcal{C}(\overline{\Omega}) \text{ for all } t \in [0,T]. \tag{2.2}$$

Let $(x_m, t_m) \in \overline{\Omega_T}$ be arbitrary with $(x_m, t_m) \to (x,t)$ in $\overline{\Omega_T}$ as $m \to \infty$. We have

$$|z(x,t) - z(x_m,t_m)| \leq \underbrace{|z(x,t) - z(x,t_m)|}_{A_m} + \underbrace{|z(x,t_m) - z(x_m,t_m)|}_{B_m}.$$

Assume that $A_m \nrightarrow 0$ as $m \to \infty$. Then, there exists a subsequence of $\{A_m\}$ (also denoted by $\{A_m\}$) such that $\lim_{m \to \infty} A_m > 0$. Using this subsequence, it holds $z(\cdot, t_m) \to z(\cdot, t)$ in $L^2(\Omega)$ due to (2.2). We obtain again a subsequence (we omit the additional subscript) such that $z(y, t_m) \to z(y, t)$ as $m \to \infty$ for a.e. $y \in \Omega$. Therefore, we can choose $y_m \to x$ in $\overline{\Omega}$ such that $|z(y_m, t) - z(y_m, t_m)| \to 0$ as $m \to \infty$. It follows

$$A_m \leq \underbrace{|z(x,t) - z(y_m, t)|}_{A_m^1} + \underbrace{|z(y_m, t) - z(y_m, t_m)|}_{A_m^2} + \underbrace{|z(y_m, t_m) - z(x, t_m)|}_{A_m^3}$$

The continuity of $z(\cdot, t)$ due to (2.2) implies $A_m^1 \to 0$ as $m \to \infty$. A_m^2 converges to 0 by the construction of $\{y_m\}$. To treat the term A_m^3, we apply the Poincaré inequality from Theorem 2.3.13 with $M := \{u \in W^{1,p}(B_1(q_0)) \,|\, u(q_0) = 0\}$ and obtain

$$\|g\|_{L^p(B_1(q_0))} \leq C \|\nabla g\|_{L^p(B_1(q_0))} \tag{2.3}$$

for all $g \in M$, where $q_0 \in \mathbb{R}^n$, and $C > 0$ is independent of g and q_0. Note that, due to $g \in W^{1,p}(B_1(q_0)) \subseteq \mathcal{C}(\overline{B_1(q_0)})$, g is pointwise defined. By utilizing (2.3) and using a scaling argument, we gain a $C > 0$ such that for all $\varepsilon > 0$ and all $g \in W^{1,p}(B_\varepsilon(q_0))$ with $g(q_0) = 0$ it follows

$$\begin{aligned}
\|g\|_{\mathcal{C}(\overline{B_\varepsilon(q_0)})} = \|g(\varepsilon \cdot)\|_{\mathcal{C}(\overline{B_1(q_0)})} &\leq C \|g(\varepsilon \cdot)\|_{W^{1,p}(B_1(q_0))} \\
&\leq C \|\varepsilon \nabla g(\varepsilon \cdot)\|_{L^p(B_1(q_0))} \\
&= C \varepsilon^{\frac{p-n}{p}} \|\nabla g\|_{L^p(B_\varepsilon(q_0))}.
\end{aligned}$$

By setting $g_m(\cdot) := z(y_m, t_m) - z(\cdot, t_m)$ and $\varepsilon_m := 2|y_m - x|$, we can estimate A_m^3 in the following way (note that $g_m(y_m) = 0$):

$$A_m^3 \leq \|g_m\|_{\mathcal{C}(\overline{B_{\varepsilon_m}(y_m)})} \leq C \varepsilon_m^{\frac{p-n}{p}} \|\nabla g_m\|_{L^p(B_{\varepsilon_m}(y_m))}.$$

Since $z \in L^\infty(0, T; W^{1,p}(\Omega)) \cap H^1(0, T; L^2(\Omega))$, $\|\nabla g_m\|_{L^p(B_{\varepsilon_m}(y_m))}$ is bounded with respect to m. In conclusion, $A_m^3 \to 0$ as $m \to \infty$. Hence, we end up with a contradiction. Therefore, $A_m \to 0$ as $m \to \infty$.

The convergence $B_m \to 0$ as $m \to \infty$ can be shown as for $A_m^3 \to 0$. $\qquad\square$

Remark 2.3.15 *The inclusion in Proposition 2.3.14 is a special version of a more general compactness result in [Sim86, Corollary 5].*

2.3.3 Approximations of $W^{1,p}(\Omega)$ and $L^q(0, T; W^{1,p}(\Omega))$-functions

This subsection presents an approximation method that can be used for passing to the limit of certain variational problems. To the author's best knowledge, the main result in this subsection (Lemma 2.3.18) has not been investigated in the literature. For more details concerning the application, see step (iii) from the proof of Theorem 4.2.5.

Lemma 2.3.16 (Approximation of test functions) *Let $p > n$ and $f, \zeta \in W_+^{1,p}(\Omega)$ with $\{\zeta = 0\} \supseteq \{f = 0\}$. Furthermore, let $\{f_M\}_{M \in \mathbb{N}} \subseteq W_+^{1,p}(\Omega)$ be a sequence with $f_M \rightharpoonup f$ in $W^{1,p}(\Omega)$ as $M \to \infty$. Then, there exist a sequence $\{\zeta_M\}_{M \in \mathbb{N}} \subseteq W_+^{1,p}(\Omega)$ and constants $\nu_M > 0$, $M \in \mathbb{N}$, such that* *(i)* $\zeta_M \to \zeta$ in $W^{1,p}(\Omega)$ as $M \to \infty$,

(ii) $\zeta_M \leq \zeta$ in Ω for all $M \in \mathbb{N}$,

(iii) $\nu_M \zeta_M \leq f_M$ in Ω for all $M \in \mathbb{N}$.

Proof. We give the following proof.

- Without loss of generality, we may assume $\zeta \not\equiv 0$ on $\overline{\Omega}$. Otherwise, the statement follows directly.

- Let $\{\delta_k\}$ be a sequence with $\delta_k \to 0^+$ as $k \to \infty$ and $\delta_k > 0$. Define for every $k \in \mathbb{N}$ the approximation function $\widetilde{\zeta}_k \in W^{1,p}_+(\Omega)$ as

$$\widetilde{\zeta}_k := [\zeta - \delta_k]^+,$$

where $[\cdot]^+$ stands for $\max\{0, \cdot\}$.

- Let $0 < \alpha < 1 - \frac{n}{p}$ be a fixed constant. In the following, we use the compact embedding

$$W^{1,p}(\Omega) \hookrightarrow \mathcal{C}^{0,\alpha}(\overline{\Omega})$$

due to Theorem 2.3.4 (ii). In particular, we obtain $f, f_M, \zeta, \widetilde{\zeta}_k \in \mathcal{C}^{0,\alpha}(\overline{\Omega})$ and $f_M \to f$ in $\mathcal{C}^{0,\alpha}(\overline{\Omega})$ as $M \to \infty$.

- Set the constant R_k, $k \in \mathbb{N}$, to

$$R_k := \left(\delta_k / \|\zeta\|_{\mathcal{C}^{0,\alpha}(\overline{\Omega})} \right)^{1/\alpha} > 0.$$

- We obtain the inclusion

$$\{\widetilde{\zeta}_k = 0\} \supseteq \overline{\Omega} \cap B_{R_k}(\{\zeta = 0\}). \tag{2.4}$$

Indeed, let $x \in \overline{\Omega} \cap B_{R_k}(\{\zeta = 0\})$. Then, it follows $dist(x, \{\zeta = 0\}) < R_k$. This implies the existence of a $y \in \{\zeta = 0\}$ with $|x - y| < R_k$. Now, we can estimate as follows:

$$|\zeta(x)| = |\zeta(x) - \underbrace{\zeta(y)}_{=0}| \leq \|\zeta\|_{\mathcal{C}^{0,\alpha}(\overline{\Omega})} |x - y|^\alpha < \|\zeta\|_{\mathcal{C}^{0,\alpha}(\overline{\Omega})} (R_k)^\alpha = \delta_k.$$

We end up with $x \in \{\widetilde{\zeta}_k = 0\}$.

- Taking also the assumption $\{\zeta = 0\} \supseteq \{f = 0\}$ into account, we obtain

$$\{\widetilde{\zeta}_k = 0\} \supseteq \overline{\Omega} \cap B_{R_k}(\{\zeta = 0\}) \supseteq \overline{\Omega} \cap B_{R_k}(\{f = 0\}). \tag{2.5}$$

- Since $\zeta \not\equiv 0$, it follows from $\{\zeta = 0\} \supseteq \{f = 0\}$ that $f \not\equiv 0$. Thus, if k is large enough, we obtain $\overline{\Omega} \setminus B_{R_k}(\{f = 0\}) \neq \emptyset$. By possibly modifying the sequence $\{\delta_k\}$, we may assume, without loss of generality, $\overline{\Omega} \setminus B_{R_k}(\{f = 0\}) \neq \emptyset$ for all $k \in \mathbb{N}$.

- Let $k \in \mathbb{N}$ be arbitrary and fixed for the moment. Since $\overline{\Omega} \setminus B_{R_k}(\{f = 0\})$ is a compact set contained in $\{f > 0\}$ (note that $\{f > 0\} := \{x \in \overline{\Omega} \,|\, f(x) > 0\}$), it follows

$$\eta_k := \inf\{f(x) \,|\, x \in \overline{\Omega} \setminus B_{R_k}(\{f = 0\})\} > 0.$$

Due to $f_M \to f$ in $\mathcal{C}(\overline{\Omega})$, there exists an $\widetilde{M} \in \mathbb{N}$ such that for all $M \geq \widetilde{M}$:

$$f_M \geq \eta_k/2 \text{ in } \overline{\Omega} \setminus B_{R_k}(\{f = 0\}).$$

- Therefore, we find a strictly increasing sequence $\{M_k\} \subseteq \mathbb{N}$ such that for all $k \in \mathbb{N}$:

$$f_M \geq \eta_k/2 \text{ in } \overline{\Omega} \setminus B_{R_k}(\{f = 0\}) \text{ for all } M \geq M_k. \tag{2.6}$$

- For all $M \geq M_k$, we obtain $\widetilde{\nu}_k \widetilde{\zeta}_k \leq f_M$ in $\overline{\Omega}$ with the constant

$$\widetilde{\nu}_k := \eta_k/(2\|\zeta\|_{L^\infty(\Omega)}) > 0. \tag{2.7}$$

Indeed, for $x \in \overline{\Omega} \cap B_{R_k}(\{f = 0\})$, we get $x \in \{\widetilde{\zeta}_k = 0\}$ by (2.5) and, therefore,

$$\widetilde{\nu}_k \widetilde{\zeta}_k(x) = 0 \leq f_M(x).$$

In the case, $x \in \overline{\Omega} \setminus B_{R_k}(\{f = 0\})$, we can use (2.7) and (2.6), and estimate as follows:

$$\widetilde{\nu}_k \widetilde{\zeta}_k(x) = \frac{\eta_k}{2} \frac{\widetilde{\zeta}_k(x)}{\|\zeta\|_{L^\infty(\Omega)}} \leq f_M(x).$$

- The claim follows with $\zeta_M := 0$ and $\nu_k := 1$ for $M \in \{1, \ldots, M_1 - 1\}$ and $\zeta_M := \widetilde{\zeta}_{\delta_k}$ and $\nu_M := \widetilde{\nu}_k$ for each $M \in \{M_k, \ldots, M_{k+1} - 1\}$, $k \in \mathbb{N}$. □

Remark 2.3.17 *We remark that $\{\zeta = 0\} \supseteq \{f = 0\}$ means in the context of Lemma 2.3.16*

$$\{x \in \overline{\Omega} \,|\, \zeta(x) = 0\} \supseteq \{x \in \overline{\Omega} \,|\, f(x) = 0\}$$

by using the embedding $W^{1,p}(\Omega) \hookrightarrow \mathcal{C}(\overline{\Omega})$ $(p > n)$.

Lemma 2.3.18 (Approximation of time-dependent test functions) *Let $p > n$, $q \geq 1$ and $f, \zeta \in L^q(0, T; W^{1,p}_+(\Omega))$ with $\{\zeta(t) = 0\} \supseteq \{f(t) = 0\}$ for a.e. $t \in (0, T)$. Furthermore, let $\{f_M\}_{M \in \mathbb{N}} \subseteq L^q(0, T; W^{1,p}_+(\Omega))$ be a sequence with $f_M(t) \rightharpoonup f(t)$ in $W^{1,p}(\Omega)$ as $M \to \infty$ for a.e. $t \in (0, T)$. Then, there exist a sequence $\{\zeta_M\}_{M \in \mathbb{N}} \subseteq L^q(0, T; W^{1,p}_+(\Omega))$ and constants $\nu_{M,t} > 0$ such that* *(i)* $\zeta_M \to \zeta$ in $L^q(0, T; W^{1,p}(\Omega))$ as $M \to \infty$,

(ii) $\zeta_M \leq \zeta$ a.e. in Ω_T for all $M \in \mathbb{N}$ *(in particular* $\{\zeta_M = 0\} \supseteq \{\zeta = 0\}$*),*

(iii) $\nu_{M,t}\zeta_M(t) \leq f_M(t)$ in Ω for a.e. $t \in (0, T)$ and for all $M \in \mathbb{N}$.

If, in addition, $\zeta \leq f$ a.e. in Ω_T then condition (iii) can be refined to

(iii)' $\zeta_M \leq f_M$ a.e. in Ω_T for all $M \in \mathbb{N}$.

Proof. Let $\{\delta_k\}$ with $\delta_k \to 0^+$ as $k \to \infty$ and $\delta_k > 0$ be a sequence and $0 < \alpha < 1 - \frac{n}{p}$ be a fixed constant. As in the proof of the previous lemma, we use the compact embedding $W^{1,p}(\Omega) \hookrightarrow \mathcal{C}^{0,\alpha}(\overline{\Omega})$ in the following.

We construct the approximations $\zeta_M \in L^q(0, T; W^{1,p}_+(\Omega))$, $M \in \mathbb{N}$, as follows:

$$\zeta_M(t) := \sum_{k=1}^{M} \chi_{A_M^k}(t)[\zeta(t) - \delta_k]^+, \tag{2.8}$$

where $\chi_{A_M^k} : [0, T] \to \{0, 1\}$ is defined as the characteristic function of the measurable set A_M^k given by

$$A_M^k := \begin{cases} P_M^k \setminus \left(\bigcup_{i=k+1}^{M} P_M^i \right) & \text{if } k < M, \\ P_M^M & \text{if } k = M, \end{cases} \tag{2.9}$$

with

$$P_M^k := \Big\{ t \in [0, T] \,|\, \overline{\Omega} \setminus B_{R_k(t)}(\{f(t) = 0\}) \neq \emptyset$$
$$\text{and } f_M(t) \geq \eta_k(t)/2 \text{ in } \overline{\Omega} \setminus B_{R_k(t)}(\{f(t) = 0\}) \Big\}, \tag{2.10}$$

where the functions $R_k : [0, T] \to [0, \infty]$ and $\eta_k : [0, T] \to [0, \infty)$ are defined by

$$R_k(t) = \begin{cases} \left(\delta_k / \|\zeta(t)\|_{\mathcal{C}^{0,\alpha}(\overline{\Omega})} \right)^{1/\alpha}, & \|\zeta(t)\|_{\mathcal{C}^{0,\alpha}(\overline{\Omega})} > 0, \\ \infty, & \text{otherwise,} \end{cases}$$
$$\eta_k(t) = \inf\{f(t, x) \,|\, x \in \overline{\Omega} \setminus B_{R_k(t)}(\{f(t) = 0\})\}.$$

Note that A_M^k, $1 \leq k \leq M$, are pairwise disjoint by construction.

We are going to prove that the construction of ζ_M satisfies (i)-(iii).

(i) By the assumptions, it holds for a.e. $t \in (0, T)$:

- $f_M(t) \rightharpoonup f(t)$ in $W^{1,p}(\Omega)$ as $M \to \infty$
- $\{\zeta(t) = 0\} \supseteq \{f(t) = 0\}$

Take such a t and consider the case $\zeta(t) \not\equiv 0$. Then, $f(t) \not\equiv 0$.

Let $K \in \mathbb{N}$ be arbitrary but large enough such that $R_K(t) > 0$ is so small that we have $\overline{\Omega} \setminus B_{R_K(t)}(\{f(t) = 0\}) \neq \emptyset$.

Since $\overline{\Omega} \setminus B_{R_K(t)}(\{f(t) = 0\})$ is a compact set contained in $\{f(t) > 0\}$ and $f_M(t) \to f(t)$ in $\mathcal{C}(\overline{\Omega})$ as $M \to \infty$, we find an $\widetilde{M} \in \mathbb{N}$ such that

$$f_M(t) \geq \eta_K(t)/2 \text{ in } \overline{\Omega} \setminus B_{R_K(t)}(\{f(t) = 0\}) \text{ for all } M \geq \widetilde{M}.$$

In other words,

$$\forall K \in \mathbb{N} \text{ large enough}, \exists \widetilde{M} \geq K, \forall M \geq \widetilde{M} : t \in P_M^K. \tag{2.11}$$

A visualization of this statement is shown Figure 2.1 and, in particular, it implies (see the definition of A_M^k in (2.9)):

$$\forall M \in \mathbb{N} \text{ large}, \exists K \in \{1, \ldots, M\} : t \in P_M^K,$$
$$\forall M \in \mathbb{N} \text{ large}: t \in A_M^{k_M} \text{ with } k_M := \max\{k \in \{1, \ldots, M\} \,|\, t \in P_M^k\}. \tag{2.12}$$

Next, we will prove that the properties (2.11) and (2.12) imply $k_M \to \infty$ as $M \to \infty$.

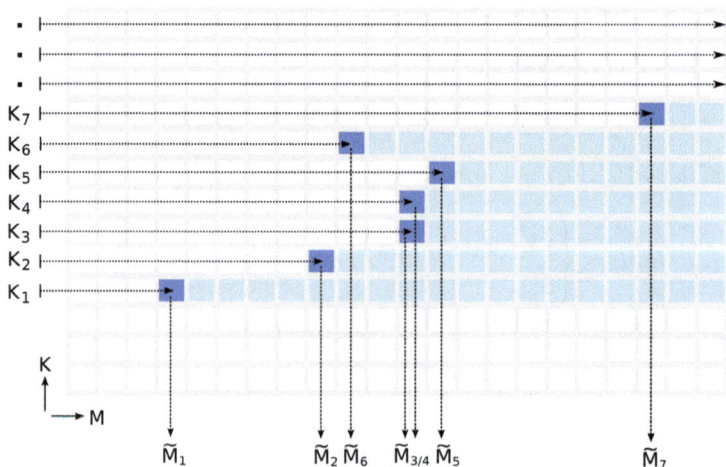

Figure 2.1: *An example to illustrate statement* (2.11): *a filled box in this matrix at position* $(M, K) \in \mathbb{N} \times \mathbb{N}$ *indicates* $t \in P_M^K$, *otherwise* $t \notin P_M^K$.

- In fact, for every large $K \in \mathbb{N}$, we find an $\widetilde{M} \geq K$ such that $t \in P_M^K$ for all $M \geq \widetilde{M}$. Then, by using (2.12), we obtain for every $M \geq \widetilde{M}$:

$$k_M = \max\{k \in \{1, \ldots, M\} \,|\, t \in P_M^k\} \geq K.$$

Consequently, $k_M \to \infty$ as $M \to \infty$.

Thus $\delta_{k_M} \to 0^+$ as $M \to \infty$.

Furthermore, $t \in A_M^{k_M}$ shows $\zeta_M(t) = [\zeta(t) - \delta_{k_M}]^+$ for every large $M \in \mathbb{N}$ by (2.8). We end up with $\zeta_M(t) \to \zeta(t)$ in $W^{1,p}(\Omega)$ as $K \to \infty$.

Taking also the estimate $\|\zeta_M(t)\|_{W^{1,p}(\Omega)} \leq \|\zeta(t)\|_{W^{1,p}(\Omega)}$ (follows from (2.8)) for all M and a.e. t into account, Lebesgue's convergence theorem shows

$$\int_0^T \|\zeta_M(t) - \zeta(t)\|_{W^{1,p}(\Omega)}^p \, dt \to 0 \text{ as } M \to \infty.$$

(ii) Property (ii) follows from (2.8) and that A_M^k, $k = 1, \ldots, M$, are pairwise disjoint.

(iii) Let t be as in (i). Since A_M^k, $k = 1, \ldots, M$, are pairwise disjoint, the definition in (2.8) implies for every $M \in \mathbb{N}$ one of the two alternatives:

- $\zeta_M(t) = 0$,
- there exists an $k \in \{1, \ldots, M\}$ such that $\zeta_M(t) = [\zeta(t) - \delta_k]^+$ and $t \in A_M^k$.

In the first case, the estimate in (iii) is fulfilled for any value $\nu_{M,t} > 0$.

In the second case, we can argue as in the proof of Lemma 2.3.16. More precisely, we obtain with the same argumentation

$$\{\zeta_M(t) = 0\} = \{[\zeta(t) - \delta_k]^+ = 0\} \supseteq \overline{\Omega} \cap B_{R_k(t)}(\{f(t) = 0\})$$

and, since $t \in P_M^k$, we obtain

$$f_M(t) \geq \eta_k(t)/2 \text{ in } \overline{\Omega} \setminus B_{R_k(t)}(\{f(t) = 0\}).$$

Now, (iii) is fulfilled with

$$\nu_{M,t} := \eta_k(t)/(2\|\zeta(t)\|_{L^\infty(\Omega)}) > 0.$$

In the case $\zeta \leq f$, we use instead of (2.10) the set

$$P_M^k := \left\{t \in [0, T] \,\Big|\, \|f_M(t) - f(t)\|_{\mathcal{C}(\overline{\Omega})} \leq \delta_k\right\}.$$

With a similar argumentation, $\{\zeta_M\}$ fulfills (i), (ii) and (iii)'. $\qquad\square$

2.3.4 Characterization of certain variational inequalities

Here, we present a technique of how one can drop a certain restriction on the space of test functions for some types of variational inequalities. To the author's best knowledge, the result in this subsection has not been investigated in the literature. In combination with the approximation technique in Subsection 2.3.3, we obtain a tool for establishing coupled variational inequalities arising from (1.1c). For more details, see step (iv) from the proof of Theorem 4.2.5.

In the following, the spelling "$\{f = 0\} \supseteq \{z = 0\}$ in an a.e. sense" means that there exists a subset $N \subseteq \Omega$ with $\mathcal{L}^n(N) = 0$ such that $(\{f = 0\} \setminus N) \supseteq (\{z = 0\} \setminus N)$.

Lemma 2.3.19 *Let $p > n$ and let $f \in L^{p/(p-1)}(\Omega; \mathbb{R}^n)$, $g \in L^1(\Omega)$ and $z \in W_+^{1,p}(\Omega)$ with $f \cdot \nabla z \geq 0$ a.e. in Ω and $\{f = 0\} \supseteq \{z = 0\}$ in an a.e. sense. Furthermore, we assume that*

$$\int_\Omega (f \cdot \nabla \zeta + g\zeta) \, dx \geq 0 \quad \text{for all } \zeta \in W_-^{1,p}(\Omega) \text{ with } \{\zeta = 0\} \supseteq \{z = 0\}. \tag{2.13}$$

Then

$$\int_\Omega (f \cdot \nabla \zeta + g\zeta) \, dx \geq \int_{\{z=0\}} [g]^+ \zeta \, dx \quad \text{for all } \zeta \in W_-^{1,p}(\Omega). \tag{2.14}$$

Proof. We assume $z \not\equiv 0$ in Ω. Let $\zeta \in W_-^{1,p}(\Omega)$ be a test function. If $\delta > 0$ is small enough we obtain

$$\overline{\Omega} \setminus B_\delta(\{z = 0\}) \neq \emptyset, \tag{2.15}$$

since $z \in \mathcal{C}(\overline{\Omega})$ due to the embedding $W^{1,p}(\Omega) \hookrightarrow \mathcal{C}(\overline{\Omega})$. Then, we can define the approximation $\zeta_\delta \in W_-^{1,p}(\Omega)$ by

$$\zeta_\delta := \max \left\{ \zeta, -z \|\zeta\|_{L^\infty} C_\delta^{-1} \right\} \tag{2.16}$$

with the constant

$$C_\delta := \inf \left\{ z(x) \,|\, x \in \overline{\Omega} \setminus B_\delta(\{z = 0\}) \right\}. \tag{2.17}$$

It holds $C_\delta > 0$ due to $z \geq 0$, property (2.15) and the continuity of z. We consider the following partition of $\overline{\Omega}$:

$$\overline{\Omega} = \Sigma_1^\delta \cup \Sigma_2^\delta \cup \Sigma_3^\delta$$

with

$$\Sigma_1^\delta := \overline{\Omega} \setminus B_\delta(\{z = 0\}),$$
$$\Sigma_2^\delta := \overline{\Omega} \cap B_\delta(\{z = 0\}) \cap \{\zeta \leq -z\|\zeta\|_{L^\infty} C_\delta^{-1}\},$$
$$\Sigma_3^\delta := \overline{\Omega} \cap B_\delta(\{z = 0\}) \cap \{\zeta > -z\|\zeta\|_{L^\infty} C_\delta^{-1}\}.$$

By the definition (2.16) with the constant (2.17), the sequence $\{\zeta_\delta\}_{\delta \in (0,1)}$ satisfies

$$\zeta_\delta(x) = \begin{cases} \zeta(x), & \text{if } x \in \Sigma_1^\delta \cup \Sigma_3^\delta, \\ -z(x)\|\zeta\|_{L^\infty} C_\delta^{-1}, & \text{if } x \in \Sigma_2^\delta, \end{cases} \tag{2.18}$$

as well as (in an a.e. sense)

$$\nabla\zeta_\delta(x) = \begin{cases} \nabla\zeta(x), & \text{if } x \in \Sigma_1^\delta \cup \Sigma_3^\delta, \\ -\nabla z(x)\|\zeta\|_{L^\infty} C_\delta^{-1}, & \text{if } x \in \Sigma_2^\delta. \end{cases} \tag{2.19}$$

From (2.16) and $\zeta \leq 0$ in Ω, we infer that $\{\zeta_\delta = 0\} \supseteq \{z = 0\}$ and

$$-\int_{\{z=0\}} [g]^+ \zeta \, dx \geq -\int_{\{z=0\}} g\zeta \, dx = -\int_{\{z=0\}} g(\zeta - \zeta_\delta) \, dx. \tag{2.20}$$

We calculate

$$\int_\Omega (f \cdot \nabla\zeta + g\zeta) \, dx - \int_{\{z=0\}} [g]^+ \zeta \, dx$$

$$= \int_\Omega (f \cdot \nabla(\zeta - \zeta_\delta) + g(\zeta - \zeta_\delta)) \, dx + \underbrace{\int_\Omega (f \cdot \nabla\zeta_\delta + g\zeta_\delta) \, dx}_{\geq 0 \text{ by (2.13)}} - \underbrace{\int_{\{z=0\}} [g]^+ \zeta \, dx}_{\text{apply (2.20)}}$$

$$\geq \int_\Omega (f \cdot \nabla(\zeta - \zeta_\delta) + g(\zeta - \zeta_\delta)) \, dx - \int_{\{z=0\}} g(\zeta - \zeta_\delta) \, dx$$

$$= \int_\Omega f \cdot \nabla(\zeta - \zeta_\delta) \, dx + \int_{\{z>0\}} g(\zeta - \zeta_\delta) \, dx$$

$$= \underbrace{\int_{\Sigma_1^\delta \cup \Sigma_3^\delta} f \cdot \nabla(\zeta - \zeta_\delta) \, dx}_{=0 \text{ by (2.19)}} + \int_{\Sigma_2^\delta} f \cdot \nabla(\zeta - \zeta_\delta) \, dx + \int_{\{z>0\}} g(\zeta - \zeta_\delta) \, dx$$

$$= \int_{\Sigma_2^\delta} f \cdot \nabla\zeta \, dx - \underbrace{\int_{\Sigma_2^\delta} f \cdot \nabla\zeta_\delta \, dx}_{\text{using (2.19)}} + \int_{\{z>0\}} g(\zeta - \zeta_\delta) \, dx$$

$$= \int_{\Sigma_2^\delta} f \cdot \nabla\zeta \, dx + \underbrace{\|\zeta\|_{L^\infty} C_\delta^{-1} \int_{\Sigma_2^\delta} f \cdot \nabla z \, dx}_{\geq 0 \text{ by assumption}} + \int_{\{z>0\}} g(\zeta - \zeta_\delta) \, dx$$

$$= \underbrace{\int_{\Sigma_2^\delta} f \cdot \nabla\zeta \, dx}_{\text{using } \{f=0\} \supseteq \{z=0\} \text{ a.e.}} + \underbrace{\int_{\{z>0\}} g(\zeta - \zeta_\delta) \, dx}_{\text{using (2.18)}}$$

$$= \int_{\Sigma_2^\delta \setminus \{z=0\}} f \cdot \nabla\zeta \, dx + \int_{\Sigma_2^\delta \setminus \{z=0\}} g(\zeta - \zeta_\delta) \, dx \tag{2.21}$$

The two terms on the right hand side can be treated as follows:

- The set $\{z = 0\}$ is closed because z is continuous. Therefore, we obtain

$$\bigcap_{\delta > 0} B_\delta(\{z = 0\}) = \{z = 0\}$$

and, consequently,

$$\bigcap_{\delta > 0} \left(B_\delta(\{z = 0\}) \setminus \{z = 0\} \right) = \emptyset.$$

The monotonicity of the measure $\mathcal{L}^n$ yields $\mathcal{L}^n\big(B_\delta(\{z = 0\}) \setminus \{z = 0\}\big) \to 0$ as $\delta \to 0^+$. This implies

$$\mathcal{L}^n\big(\Sigma_2^\delta \setminus \{z = 0\}\big) \le \mathcal{L}^n\big(B_\delta(\{z = 0\}) \setminus \{z = 0\}\big) \to 0 \qquad (2.22)$$

and we end up with

$$\int_{\Sigma_2^\delta \setminus \{z=0\}} f \cdot \nabla \zeta \, \mathrm{d}x \to 0 \qquad (2.23)$$

as $\delta \to 0^+$.

- Since $\zeta \le 0$ and $z \ge 0$, it follows from the definition of ζ_δ in (2.16) that $\zeta \le \zeta_\delta \le 0$. Therefore, ζ_δ is uniformly bounded in the $L^\infty(\Omega)$ norm and we can argue as follows by taking (2.22) into account:

$$\left| \int_{\Sigma_2^\delta \setminus \{z=0\}} g(\zeta - \zeta_\delta) \, \mathrm{d}x \right| \le C \int_{\Sigma_2^\delta \setminus \{z=0\}} |g| \, \mathrm{d}x \to 0 \qquad (2.24)$$

as $\delta \to 0^+$.

From (2.21) and (2.23) and (2.24), we infer the claim (2.14). $\square$

Remark 2.3.20 *Let us remark that we suppose* $\{f = 0\} \supseteq \{z = 0\}$ *only in an a.e. sense in Lemma 2.3.19.*

2.4 Shrinking sets and admissible subsets

The aim of this section is to prove covering and representation results for shrinking sets as well as to introduce local Sobolev spaces on shrinking sets. We call a space-time subset $G \subseteq \overline{\Omega_T}$ shrinking if G is relatively open in $\overline{\Omega_T}$ and $G(s) \subseteq G(t)$ holds whenever $0 \le t \le s \le T$. Here, the t-cut of G at time $t \in [0, T]$, i.e., $G \cap (\overline{\Omega} \times \{t\})$, is denoted by $G(t) := \{x \in \overline{\Omega} \,|\, (x, t) \in G\}$.

2.4.1 Covering and representation results

In the following, we are going to study shrinking sets G. They will appear in the analysis of the complete damage systems in Chapter 6 and Chapter 7 as subsets of $\overline{\Omega_T}$ where the damage is not complete. Due to the possibly bad smoothness property of G, we will need to represent certain parts of G as a countable union of Lipschitz domains. In this context, it is convenient to introduce the notion of *fine representation* and of *admissible subset*.

Definition 2.4.1 (Fine representation) *Let $H \subseteq \overline{\Omega}$ be a relatively open subset. We call a countable family $\{U_k\}$ of open sets $U_k \subset\subset H$ a fine representation for H if for every $x \in H$ there exist an open set $U \subseteq \mathbb{R}^n$ with $x \in U$ and an $k \in \mathbb{N}$ such that $(U \cap \Omega) \subseteq U_k$.*

Remark 2.4.2 *See Figure 2.2 for an example. Note that $H \cap \partial\Omega$ is not covered by $\{U_k\}$.*

For a given relatively open subset of $\overline{\Omega}$, we will be interested in the subsets where every path-connected component is connected to the Dirichlet boundary Γ_{D}.

Definition 2.4.3 (Admissible subsets of $\overline{\Omega}$ with respect to Γ_{D})

(i) *Let $F \subseteq \overline{\Omega}$ be a relatively open subset and*

$$P_F(x) := \big\{ y \in F \,|\, x \text{ and } y \text{ are connected by a path in } F \big\}$$

for $x \in F$. We say that F is admissible with respect to the Dirichlet boundary Γ_{D} if for every $x \in F$ the condition

$$\mathcal{H}^{n-1}(P_F(x) \cap \Gamma_{\mathrm{D}}) > 0$$

is fulfilled. Furthermore, $\mathfrak{A}_{\Gamma_{\mathrm{D}}}(F)$ denotes the maximal admissible subset of F with respect to Γ_{D}, i.e.,

$$\mathfrak{A}_{\Gamma_{\mathrm{D}}}(F) := \bigcup \{ G \subseteq F \,|\, G \text{ is admissible with respect to } \Gamma_{\mathrm{D}} \}.$$

(ii) *For a relatively open subset $F \subseteq \overline{\Omega_T}$, the set $\mathfrak{A}_{\Gamma_{\mathrm{D}}}(F)$ is given by $(\mathfrak{A}_{\Gamma_{\mathrm{D}}}(F))(t) := \mathfrak{A}_{\Gamma_{\mathrm{D}}}(F(t))$.*

In the remaining part of this subsection, we are going to prove certain covering and representation results.

Lemma 2.4.4 (Finite covering) *Let $G \subseteq \overline{\Omega_T}$ be a relatively open subset and the sequence $\{t_m\}$ containing T be dense in $[0,T]$. Furthermore, let $\{U_k^m\}_{k \in \mathbb{N}}$ be a fine representation for $G(t_m)$ for every $m \in \mathbb{N}$. Then, for every compact set $K \subseteq G$ there exist a finite set $I \subseteq \mathbb{N}$ and values $m_k \in \mathbb{N}$, $k \in I$, such that $K \cap \Omega_T \subseteq \bigcup_{k \in I} (U_k^{m_k} \times (0, t_{m_k}))$.*

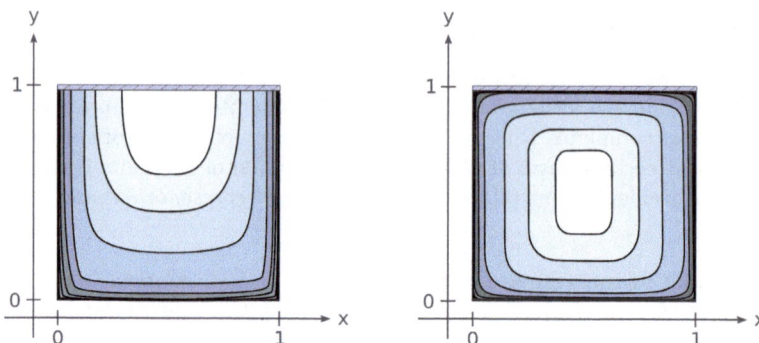

Figure 2.2: *Left: a fine representation for the relatively open subset $H = (0,1) \times (0,1]$ of $\overline{\Omega} = [0,1] \times [0,1]$; Right: not a fine representation for H.*

Proof. To every element $p = (x,t) \in K$, we will construct a neighborhood $\Theta_p \subseteq \overline{\Omega_T}$ of p in the subspace topology of $\overline{\Omega_T}$ such that there exists $k, m \in \mathbb{N}$ with $\Theta_p \cap \Omega_T \subseteq U_k^m \times (0, t_m)$. Then the claim follows by the Heine-Borel theorem.

Indeed, to every $p = (x,t) \in K$ there exists an $\varepsilon > 0$ such that $B_\varepsilon(p) \cap \overline{\Omega_T} \subseteq G$ since $G \subseteq \overline{\Omega_T}$ is relatively open. Therefore, if $t < T$, $(x, t_m) \in G$ for all $m \in \mathbb{N}$ such that $t < t_m < t + \varepsilon$. This implies $(x, t_m) \in G \cap (\overline{\Omega} \times \{t_m\}) = G(t_m) \times \{t_m\}$. Then, we find $p \in G(t_m) \times J$ with $J = [0, t_m)$. In the case $t = T$, it holds $p \in G(T) \times J$ with $J = [0, T]$. Since $\{U_k^m\}_{k \in \mathbb{N}}$ is a fine representation of $G(t_m)$, let $\delta > 0$ such that $B_\delta(x) \cap \Omega \subseteq U_k^m$ for some $k \in \mathbb{N}$. Finally, $\Theta_p := (B_\delta(x) \cap \overline{\Omega}) \times J$ is the required neighborhood of p. □

Lemma 2.4.5 (Partition of unity property) *Let G, $\{t_m\}$ and $\{U_k^m\}$ be as in Lemma 2.4.4. Then, for every compact subset $K \subseteq G$ there exist a finite set $I \subseteq \mathbb{N}$, values $m_k \in \mathbb{N}$, $k \in I$ and functions $\psi_k \in \mathbb{C}^\infty(\overline{\Omega_T})$, $k \in I$, such that*

(i) $K \cap \Omega_T \subseteq \bigcup_{k \in I} \left(U_k^{m_k} \times (0, t_{m_k}) \right),$

(ii) $\operatorname{supp}(\psi_k) \subseteq \overline{U_k^{m_k}} \times [0, t_{m_k}],$

(iii) $\sum_{k \in I} \psi_k \equiv 1$ *on K.*

Proof. We extend the family of open sets $\{V_k^m\}$ given by $V_k^m := U_k^m \times (0, t_{m_k})$ in the following way. Define

$$\mathcal{P} := \left\{ \{W_k^m\} \mid W_k^m \subseteq \mathbb{R}^{n+1} \text{ is open with } W_k^m \cap \Omega_T = U_k^m \times (0, t_{m_k}) \right\}.$$

We see that $\mathcal{P}$ is non-empty and every totally ordered subset of $\mathcal{P}$ has an upper bound with respect to the "$\leq$" ordering defined by

$$\{W_k^m\} \leq \{\widetilde{W}_k^m\} \Leftrightarrow W_k^m \subseteq \widetilde{W}_k^m \text{ for all } k, m \in \mathbb{N}.$$

By Zorn's lemma, we find a maximal element $\{\widetilde{V}_k^m\}$. It holds

$$G \subseteq \bigcup_{k,m \in \mathbb{N}} \widetilde{V}_k^m. \tag{2.25}$$

Assume that this condition fails. Because of $G \cap \Omega_T = \bigcup_{k,m \in \mathbb{N}} V_k^m$, there exists a $p = (x,t) \in G \cap \partial(\Omega_T)$ with $p \notin \bigcup_{k,m \in \mathbb{N}} \widetilde{V}_k^m$.

Let us consider the case $t < T$. Since $F \subseteq \overline{\Omega_T}$ is relatively open, we find an $m_0 \in \mathbb{N}$ with $x \in G(t_{m_0})$ and $t_{m_0} > t$. By the fine representation property of $\{U_k^{m_0}\}_{k \in \mathbb{N}}$ for $G(t_{m_0})$, we find an open set $U \subseteq \mathbb{R}^n$ with $x \in U$ and $k_0 \in \mathbb{N}$ such that $U \cap \Omega \subseteq U_{k_0}^{m_0}$.

The family $\{\widetilde{W}_k^m\}$ given by

$$\widetilde{W}_k^m := \begin{cases} \widetilde{V}_k^m \cup U \times (-\infty, t_{m_0}) & \text{if } k = k_0 \text{ and } m = m_0, \\ \widetilde{V}_k^m & \text{else}, \end{cases}$$

satisfies $\{\widetilde{W}_k^m\} \in \mathcal{P}$ and $p \in \bigcup_{k,m \in \mathbb{N}} \widetilde{W}_k^m$ which contradicts the maximality property of $\{\widetilde{V}_k^m\}$.

In the case $t = T$, we also find $k_0, m_0 \in \mathbb{N}$ and an open set $U \subseteq \mathbb{R}^n$ with $x \in U$ such that $U \cap \Omega \subseteq U_{k_0}^{m_0}$ and $t_{m_0} = T$. The family $\{\widetilde{W}_k^m\}$ given by

$$\widetilde{W}_k^m := \begin{cases} \widetilde{V}_k^m \cup U \times \mathbb{R} & \text{if } k = k_0 \text{ and } m = m_0, \\ \widetilde{V}_k^m & \text{else}, \end{cases}$$

also contradicts the maximality of $\{\widetilde{V}_k^m\}$. Therefore, (2.25) is proven.

Heine-Borel theorem yields

$$K \subseteq \bigcup_{k \in I} \widetilde{V}_k^{m_k}$$

for a finite set $I \subseteq \mathbb{N}$ and values $m_k \in \mathbb{N}$, $k \in I$. Together with a partition of unity argument, we get functions $\psi_k \in \mathcal{C}^\infty(\overline{\Omega_T})$ such that (i)-(iii) hold. $\qquad\square$

If a relatively open set $H \subseteq \overline{\Omega}$ is admissible with respect to Γ_D we can construct a fine representation for H with Lipschitz domains in the following sense.

Lemma 2.4.6 (Lipschitz representation of admissible sets) *Let $H \subseteq \overline{\Omega}$ be relatively open and admissible with respect to Γ_D. Then, there exists a fine representation $\{U_m\}$ for H such that*

(i) U_m is a Lipschitz domain for all $m \in N$,

(ii) $\mathcal{H}^{n-1}(\partial U_m \cap \Gamma_D) > 0$ for all $m \in N$.

Proof. We will sketch a possible construction for reader's convenience.

We assume WLOG that H is path-connected because H can only have at most countably many path-connected components and for each component we can apply the construction below.

Let us choose a reference point $x_0 \in \Gamma_D \cap H$ with the property

$$\mathcal{H}^{n-1}\left(\partial\big(B_\varepsilon(x_0) \cap \Omega\big) \cap \Gamma_D\right) > 0 \text{ for all } \varepsilon > 0, \tag{2.26}$$

which is possible since $\mathcal{H}^{n-1}(\Gamma_D \cap H) > 0$. The relatively open subset $D_m \subseteq \overline{\Omega}$ for $m \in \mathbb{N}$ is defined as

$$D_m := H \setminus \overline{B_{1/m}(\overline{\Omega} \setminus H)}.$$

If m is large enough we have $x_0 \in D_m$ since $H \subseteq \overline{\Omega}$ is relatively open. We define

$$D'_m := \{x \in D_m \mid x \text{ is path-connected to } x_0 \text{ in } D_m\}.$$

Hence, we obtain an $\varepsilon > 0$ such that $B_\varepsilon(x_0) \cap \overline{\Omega} \subseteq D'_m$ since D'_m is relatively open in $\overline{\Omega}$. In combination with (2.26), this yields $\mathcal{H}^{n-1}(\partial D'_m \cap \Gamma_D) > 0$. Because of $D'_m \subset\subset H$, there exists a Lipschitz domain $U_m \subseteq \Omega$ with $D'_m \subseteq \overline{U_m} \subseteq H$ (e.g. the part of the boundary $\partial U_m \setminus \partial \Omega$ of U_m can be constructed by polygons such that ∂U_m fulfills the Lipschitz boundary condition, see Figure 2.3). The family $\{U_m\}$ satisfies all the desired properties.

$\square$

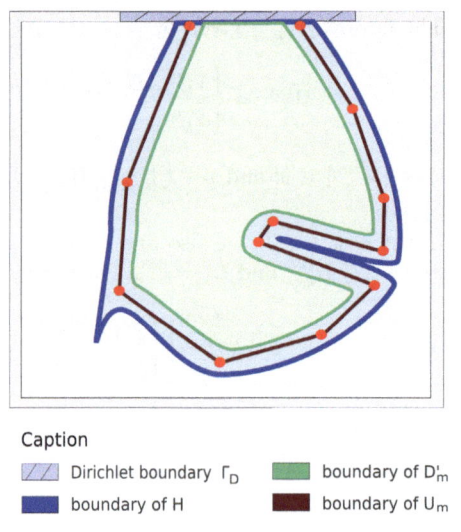

Caption

░░ Dirichlet boundary Γ_D ▇ boundary of D'_m

▇ boundary of H   ▇ boundary of U_m

Figure 2.3: *Visualization of the construction of U_m in 2D.*

Corollary 2.4.7 *Let $G \subseteq \overline{\Omega_T}$ be a shrinking set where $G(t)$ is admissible with respect to Γ_D for all $t \in [0, T]$. Furthermore, let $\{t_m\} \subseteq [0, T]$ be a dense sequence containing T.*

Then, there exists a countable family $\{U_k^m\}_{k \in \mathbb{N}}$ of Lipschitz domains $U_k^m \subset\subset G(t_m)$ for each $m \in \mathbb{N}$ such that

(i) $\mathcal{H}^{n-1}(\partial U_k^m \cap \Gamma_D) > 0$ for all $m \in \mathbb{N}$,

(ii) $\{U_k^m\}_{k \in \mathbb{N}}$ is a fine representation for $G(t_m)$ for all $m \in \mathbb{N}$,

(iii) $G = \bigcup_{m=1}^{\infty} G(t_m) \times [0, t_m].$

2.4.2 Local Sobolev spaces on shrinking sets

Given a shrinking set G, the space of local Sobolev functions on G which are of $L^2(H^q)$-type will be introduce. This space will appear in weak formulations of the complete damage approach in Chapter 6 and Chapter 7.

Definition 2.4.8 (Space-time local Sobolev functions) *Let $N \in \mathbb{N}$, $q \geq 1$ and $G \subseteq \overline{\Omega_T}$ be a shrinking set. Define*

$$L_t^2 H_{x,\mathrm{loc}}^q(G; \mathbb{R}^N) := \left\{ v : G \to \mathbb{R}^N \mid \forall t \in (0,T], \forall U \subset\subset G(t) \text{ open} : \right. \tag{2.27}$$
$$\left. v|_{U \times (0,t)} \in L^2(0,t; H^q(U; \mathbb{R}^N)) \right\}.$$

As usual, we set $L_t^2 H_{x,\mathrm{loc}}^q(G) := L_t^2 H_{x,\mathrm{loc}}^q(G; \mathbb{R})$.

Remark 2.4.9 *(i) Note that we do not demand that G is an open set.*

(ii) Applying Lemma 2.4.5 (partition of unity) shows that $L_t^2 H_{x,\mathrm{loc}}^0(G; \mathbb{R}^N)$ coincides with $L_{\mathrm{loc}}^2(G; \mathbb{R}^N)$, where $L_{\mathrm{loc}}^2(G; \mathbb{R}^N)$ denotes the classical local L^2-Lebesgue space on G given by

$$L_{\mathrm{loc}}^2(G; \mathbb{R}^N) := \left\{ v : G \to \mathbb{R}^N \mid v|_V \in L^2(V; \mathbb{R}^N) \text{ for all open } V \subset\subset G \right\}.$$

This can be seen as follows. The inclusion

$$L_{\mathrm{loc}}^2(G; \mathbb{R}^N) \subseteq L_t^2 H_{x,\mathrm{loc}}^0(G; \mathbb{R}^N)$$

follows from Definition 2.4.8 since $(U \times (0,t)) \subset\subset G$ for every $U \subset\subset G(t)$. Now, let $v \in L_t^2 H_{x,\mathrm{loc}}^0(G; \mathbb{R}^N)$ and $V \subset\subset G$ be arbitrary. Furthermore, let $\{U_k^m\}$ be as in Corollary 2.4.7.

By Lemma 2.4.5 applied to $K = \overline{V}$, we find a finite covering $\{U_k^{m_k} \times (0, t_{m_k})\}_{k \in I}$ of $K \cap \Omega_T$. For each $k \in I$, it holds $v|_{U_k^{m_k} \times (0, t_{m_k})} \in L^2(0, t_{m_k}; H^0(U_k^{m_k}; \mathbb{R}^N))$ by (2.27). Using the partition of unity yields $v|_V \in L^2(V; \mathbb{R}^N)$. Thus $v \in L_{\mathrm{loc}}^2(G; \mathbb{R}^N)$.

(iii) At fixed time points $t \in (0,T)$, we find $v(t) \in H_{\mathrm{loc}}^q(G(t); \mathbb{R}^N)$.

Given $v \in L_t^2 H_{x,\mathrm{loc}}^q(G; \mathbb{R}^N)$, we say that $v = b$ on $(\Gamma_D)_T \cap G$ if for every $t \in (0,T)$ and every open set $U \subset\subset G(t)$ with Lipschitz boundary

$$\widetilde{v}(s) = b(s) \text{ on } \partial U \cap \Gamma_D \text{ in the sense of traces for a.e. } s \in (0,t), \tag{2.28}$$

is fulfilled with $\widetilde{v} := v|_{U \times (0,t)} \in L^2(0,t; H^1(U; \mathbb{R}^N))$.

We write ∇v for the weak derivative with respect to the spatial variable as well as $\epsilon(v) := \frac{1}{2}(\nabla v + (\nabla v)^{\mathrm{t}})$ for its symmetric part. The precise definition and characterization of ∇v can be found in the following proposition.

Proposition 2.4.10 *Let $G \subseteq \overline{\Omega_T}$ be a shrinking subset and let $\{t_m\}$ and $\{U_k^m\}$ be as in Lemma 2.4.4. Furthermore, let $v : G \to \mathbb{R}^N$ be a function.*

(a) The following statements are equivalent:

(i) $v \in L_t^2 H_{x,\text{loc}}^1(G; \mathbb{R}^N)$

(ii) $v|_{U_k^m \times (0,t_m)} \in L^2(0, t_m; H^1(U_k^m; \mathbb{R}^N))$ *for all* $k, m \in \mathbb{N}$

(iii) $v \in L_{\text{loc}}^2(G; \mathbb{R}^N)$ *and there exists a function* $g \in L_{\text{loc}}^2(G; \mathbb{R}^{N \times n})$ *such that*

$$\int_G v \cdot \text{div}(\zeta) \, dx \, dt = -\int_G g : \zeta \, dx \, dt \qquad (2.29)$$

for all $\zeta \in \mathcal{C}_c^\infty(\text{int}(G); \mathbb{R}^{N \times n})$

If one of these conditions is satisfied we write $\nabla v := g$ *and* $\epsilon(v) := \frac{1}{2}(\nabla v + (\nabla v)^t)$.

(b) Assume that each U_k^m has a Lipschitz boundary. Then the following statements are equivalent:

(i) $v = b$ *on the boundary* $D_T \cap G$

(ii) for every $k, m \in \mathbb{N}$, *condition (2.28) is satisfied for* $U = U_k^m$ *and* $t = t_m$

Proof.

(a) (i)$\Longrightarrow$(ii) and (iii)$\Longrightarrow$(i) are trivial.

(ii)$\Longrightarrow$(iii): Let the function $\widehat{g} : G \to \mathbb{R}^{N \times n}$ be $\mathcal{L}^{n+1}$-a.e. defined as follows. For each $k, m \in \mathbb{N}$, we set $\widehat{g}|_{U_k^m} := \widehat{g}_k^m$ where $\widehat{g}_k^m \in L^2(U_k^m \times (0, t_m); \mathbb{R}^{N \times n})$ is the weak derivative of $v|_{U_k^m \times (0,t_m)}$. The function $\widehat{g}$ is well-defined on $G \cap \Omega_T$ since

$$G \cap \Omega_T = \bigcup_{k,m \in \mathbb{N}} U_k^m \times (0, t_m)$$

and $\widehat{g}_{k_1}^{m_1} = \widehat{g}_{k_2}^{m_2}$ on $U_{k_1}^{m_1} \times (0, t_{m_1}) \cap U_{k_2}^{m_2} \times (0, t_{m_2})$ for all $k_1, k_2, m_1, m_2 \in \mathbb{N}$ in an $\mathcal{L}^{n+1}$-a.e. sense. Let $t \in (0, T]$ and $U \subset\subset G(t)$ be open. By Lemma 2.4.4, $U \times (0, t)$ can be covered by finitely many sets $U_k^m \times (0, t_m)$. In particular, $\widehat{g}|_{U \times (0,t)} \in L^2(0, t; L^2(U; \mathbb{R}^{N \times n}))$. Thus $\widehat{g} \in L_{\text{loc}}^2(G; \mathbb{R}^{N \times n})$.

Let $\zeta \in \mathcal{C}_c^\infty(\text{int}(G); \mathbb{R}^{N \times n})$. Applying Lemma 2.4.4 again, there exists a finite set $I \subseteq \mathbb{N}$ such that $\text{supp}(\zeta) \subseteq \bigcup_{k \in I} U_k^{m_k} \times (0, t_{m_k}) =: U$. By a partition of unity argument over U, (2.29) holds for $g = \widehat{g}$.

(b) (ii)$\Longrightarrow$(i): Let $t \in (0, T)$ and $U \subset\subset G(t)$ be an arbitrary open subset. By Lemma 2.4.4, we find a finite set $I \subseteq \mathbb{N}$ such that $U \subseteq \bigcup_{k \in I} U_k^{m_k}$ and $t_{m_k} \geq t$. The claim follows. $\square$

CHAPTER 3

PDE modeling and thermodynamic consistency

This chapter is devoted to phase-field models for damage processes and phase separation in elastically stressed alloys. The associated PDE systems for both models have been separately derived from balance laws and constitutive equations by M. E. Gurtin [Gur96] for phase separation as well as by M. Frémond and B. Nedjar [FN96] for damage processes, respectively.

In Section 3.1 and Section 3.2, by reviewing concepts from [FN96, Gur96, Gar00], the physical background of both models is explained, classical PDE formulations are derived and thermodynamic consistencies are shown. For the considered damage model, several non-smooth restrictions such as uni-directionality lead to a differential inclusion formulation for the evolutionary law. However, Section 3.2 only covers partial damage processes. For a more elaborate modeling of damage processes, a possible complete failure of material's integrity due to high exposures needs to be taken into account. To this end, a new modeling approach is developed in Section 3.3 to handle the complete damage case.

Finally, Section 3.4 combines the ideas from Section 3.1 and Section 3.2 as well as Section 3.3 to introduce a unifying model for phase separation and damage. We distinguish between the partial and the complete damage case.

3.1   Phase separation and coarsening processes

Multi-component alloys alter their microstructure over time by minimizing their free energy. As long as the temperature is above a critical value, the alloy prefers one homogeneous phase. The situation rapidly changes when the medium is cooled down below the critical temperature. Then, the chemical potential function yields two or more local minima (e.g. double well potential) and a phase separation process is initiated (also called *spinodal decomposition*). Because, under this condition, it is energetically favorable for the alloy to decompose into several distinguished phases in order to minimize the bulk chemical energy. Each phase in this mixture has a fixed ratio of its chemical constituents and corresponds to a local minimum of the chemical energy density function.

As pointed out in [Gar00] (see also the references therein), the phase separation process yields the formation of fine-grained structures in the material on a very fast time scale while the long term evolution is dominated by a minimization of the interfacial free energy, i.e., a minimization of the surface area of the phases under the conservation of mass. This eventually results in coarsening processes of the microstructures. The shape of the phases become round and the number of connected regions of the phases (also called *particles*) reduces. Figure 3.1 illustrates the coarsening phenomenon for a binary alloy.

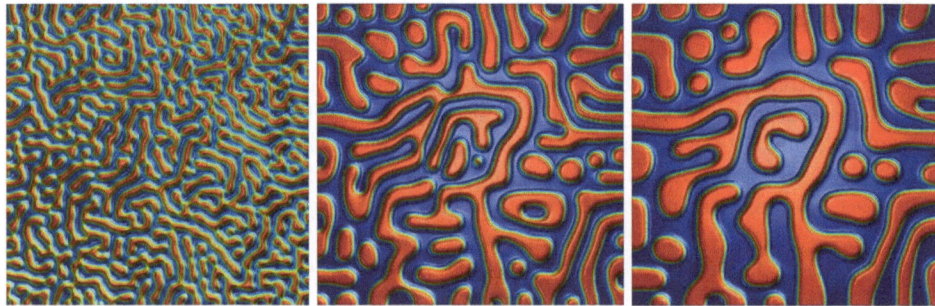

Figure 3.1: *Numerical simulation of coarsening processes in binary alloys (conducted by Rüdiger Müller from the Weierstrass Institute in Berlin); Left: fine-grained structures at the beginning; Middle, right: system at later times.*

Phase separation in alloys can mathematically be modeled by Cahn-Hilliard equations [Cah61] or by Allen-Cahn type equations [AC79] (see also [Gur96]). The former model type describes mass preserving and the latter one (usually mass non-preserving) phase-transition processes in multi-component alloys. Analytical investigations for Cahn-Hilliard equations can be found in [Ell89, CGPGS12, BP05] and for Allen-Cahn equations in [CGPGS10, CP08, BW05], respectively. Numerical simulations have been carried out in [Ell89, BB99].

Physical observations and numerical simulations involving anisotropic elastic energies and different lattice spacings reveal that elastic properties significantly influence the developing shapes of the particles (see [GW05]). To account for elasticity, the Cahn-Hilliard equations have been coupled with the momentum balance equation. The resulting PDE system is sometimes called Cahn-Larché system, cf. [LC82, Gur96] for modeling aspects. Analysis for various types of the Cahn-Larché system has been performed in [CMP00, BCD⁺02, Gar00, Gar05a, Gar05b]. For numerical results and simulations, we refer to [GRW01, Wei01, Mer05, BM10]. Gurtin pointed out in [Gur96] that certain generalized Cahn-Hilliard equations coupled with elasticity can be derived by a mass balance law, a momentum balance law and by introducing a new balance law for so-called *microforces* together with constitutive equations which are consistent with the second law of thermodynamics.

In the following presentation, we sketch the idea in [Gur96] for binary alloys. We assume small strains, i.e., $e = \varepsilon(u) := \frac{1}{2}(\nabla u + (\nabla u)^t)$, and constant temperature. Non-isothermal Cahn-Hilliard equations are treated in [AP92, KN94, GR07] (without elasticity). Since we assume only two species in the alloy, the mixture can be described by a scalar-valued variable which is associated with the concentration difference of the two components. Furthermore, we assume that the mechanical equilibrium for the elastic deformations is always attained during the phase separation process, i.e., the momentum balance becomes a quasi-static equilibrium equation.

The mentioned balance laws considered in [Gur96] read as

$$\partial_t c = -\text{div} h + m \tag{3.1a}$$

$$\text{div}\xi + \pi + \gamma = 0, \tag{3.1b}$$

$$\text{div}\sigma = l. \tag{3.1c}$$

Equation (3.1a) is referred to as the mass balance law, (3.1b) is the balance law for microforces and (3.1c) specifies the balance law for macroscopic forces. Furthermore, h denotes the mass flux, m the external mass supply, ξ the microstress, π the internal microforce, γ the external microforce, σ the Cauchy stress tensor and l the volume force. To shorten the presentation, we assume $m = \gamma = l = 0$. Note that we do not specify boundary conditions at the moment.

Let ψ denote the free energy density and consider the subsequent constitutive equations:

$$\psi = \psi(c, \nabla c, e, \mu, \nabla\mu), \quad \sigma = \sigma(c, \nabla c, e, \mu, \nabla\mu), \quad h = h(c, \nabla c, e, \mu, \nabla\mu),$$
$$\xi = \xi(c, \nabla c, e, \mu, \nabla\mu), \quad \pi = \pi(c, \nabla c, e, \mu, \nabla\mu).$$

Here, μ is referred to as the chemical potential. By requesting the following version of the local dissipation inequality (which is equivalent to the second law of thermodynamics for isothermal systems, cf. [Gur96]):

$$\partial_t\psi - \sigma : \partial_t e + (\pi - \mu)\partial_t c - \xi \cdot \partial_t\nabla c + h \cdot \nabla\mu \leq 0,$$

and by using the chain rule, we obtain the inequality (set $Z := (c, \nabla c, e, \mu, \nabla\mu)$)

$$[\partial_c\psi(Z) + \pi(Z) - \mu]\partial_t c + [\partial_{\nabla c}\psi(Z) - \xi(Z)] \cdot \partial_t\nabla c + [\partial_e\psi(Z) - \sigma(Z)] : \partial_t e$$
$$+ [\partial_\mu\psi(Z)]\partial_t\mu + [\partial_{\nabla\mu}\psi(Z)] \cdot \partial_t\nabla\mu + h(Z) \cdot \nabla\mu \leq 0$$

holding for all sufficiently smooth solutions Z of the above system. Therefore, the following constitutive relations fulfills the dissipation inequality:

$$\partial_c \psi(Z) = \mu - \pi(Z), \qquad \partial_{\nabla c} \psi(Z) = \xi(Z), \qquad \partial_e \psi(Z) = \sigma(Z), \tag{3.2a}$$
$$\partial_\mu \psi(Z) = 0, \qquad \partial_{\nabla \mu} \psi(Z) = 0, \qquad h(Z) = -\mathbb{M}(Z)\nabla\mu, \tag{3.2b}$$

where $\mathbb{M}$ indicates the diffusion mobility. Using these relations and the microforce balance equation (3.1b), we derive an explicit formula for the chemical potential:

$$\mu = -\operatorname{div}(\partial_{\nabla c}\psi(c, \nabla c, e)) + \partial_c \psi(c, \nabla c, e).$$

From now on, we assume N constituents in the alloy, i.e., the concentration function c maps from Ω_T to $\mathbb{R}^N$ and is subject to the constraints

$$c_1 + c_2 + \ldots + c_N = 1 \text{ in } \Omega_T, \tag{3.3a}$$
$$0 \le c_k \le 1 \text{ in } \Omega_T \text{ for every } k = 1, \ldots, N. \tag{3.3b}$$

In the following PDE model, the second constraint (3.3b) can only be achieved by logarithmic potentials or by a subdifferential inclusion formulation of system (3.1)-(3.2), see [BCD$^+$02] and [CGPGS12] for generalized Cahn-Hilliard equations which are highly nonlinearly coupled.

To obtain explicit PDEs for the Cahn-Larché system, we consider the following free energy density of Ginzburg-Landau type (cf. [Gar00]):

$$\psi(c, \nabla c, e) = \frac{1}{2}\nabla c : \Gamma \nabla c + W^{\text{el}}(c, e) + W^{\text{ch}}(c), \tag{3.4}$$

where W^{el} denotes the elastic energy density, W^{ch} the chemical energy density and Γ an energy gradient tensor. The gradient term $\frac{1}{2}\nabla c : \Gamma \nabla c$ corresponds to the diffuse surface energy.

An important example of an elastic energy density W^{el} for the Cahn-Larché system is given by

$$W^{\text{el}}(c, e) = \frac{1}{2}(e - e^\star(c)) : \mathbf{C}(c)(e - e^\star(c)).$$

The 4th order stiffness tensor denoted by $\mathbf{C}$ may depend on the concentration, is assumed to be positive definite and the components satisfy the symmetry condition

$$\mathbf{C}_{ijkl} = \mathbf{C}_{jikl} = \mathbf{C}_{klij}. \tag{3.5}$$

The stress free strain is denoted by $e^\star$ and takes the mismatch in the microscopic structure of the material into account [DM00, Gar00] and, in general, it depends on the concentration vector c. If a linear relation between c and $e^\star$ exists, i.e.,

$$e^\star(c) = \sum_{k=1}^{N} c_j e^\star(e_k), \text{ where } e_1, \ldots, e_N \text{ are the standard basis vectors,} \tag{3.6}$$

it is said that the material satisfies Vergard's law [DA91] which is assumed in [Gar00].

As already mentioned, phase separation occurs at sufficiently low temperatures when the chemical potential function W^{ch} becomes a double (or multi) well potential. In this regime, polynomial and logarithmic ansatz functions are deployed. Typical examples are 4th order polynomial potentials of the type

$$W^{\text{ch,pol}}(c) = \kappa_1(1 - (\kappa_2 c - \kappa_3)^2)^2, \qquad \kappa_1 > 0,\ \kappa_2, \kappa_3 \in \mathbb{R}, \tag{3.7}$$

whereas logarithmic potentials are given by

$$W^{\text{ch,log}}(c) = \theta \sum_{k=1}^{N} c_k \log(c_k) + \frac{1}{2} c \cdot Ac, \qquad \theta > 0,\ A \in \mathbb{R}^{n \times n}_{\text{sym}}. \tag{3.8}$$

This kind of energy densities can be derived from a mean field theory (see [Gar00]). The first term indicates the logarithmic entropy term and the second term the pairwise interaction term.

Completing the balance laws (3.1), the constitutive relations (3.2) and (3.4) with initial-boundary conditions, we obtain the following PDE system:

A pair of functions (c, u) with $c : \overline{\Omega_T} \to \mathbb{R}^N$ and $u : \overline{\Omega_T} \to \mathbb{R}^n$ is called a classical solution for the Cahn-Larché system to the initial-boundary data (c^0, b) if

$$\partial_t c = \text{div}(\mathbb{M}(Z)\nabla\mu) \qquad\qquad \text{in } \Omega_T, \tag{3.9a}$$
$$\mu = -\text{div}(\Gamma\nabla c) + \partial_c W^{\text{el}}(c, \epsilon(u)) + \partial_c W^{\text{ch}}(c) \qquad \text{in } \Omega_T, \tag{3.9b}$$
$$0 = \text{div}(\partial_e W^{\text{el}}(c, \epsilon(u))) \qquad\qquad \text{in } \Omega_T \tag{3.9c}$$

and the initial-boundary conditions

$$c(0) = c^0 \qquad\qquad \text{in } \Omega, \tag{3.10a}$$
$$u = b \qquad\qquad \text{on } (\Gamma_{\text{D}})_T, \tag{3.10b}$$
$$\partial_e W^{\text{el}} \cdot \nu = 0 \qquad\qquad \text{on } (\Gamma_{\text{N}})_T, \tag{3.10c}$$
$$\mathbb{M}(Z)\nabla\mu \cdot \nu = 0 \qquad\qquad \text{on } (\partial\Omega)_T, \tag{3.10d}$$
$$\Gamma\nabla c \cdot \nu = 0 \qquad\qquad \text{on } (\partial\Omega)_T \tag{3.10e}$$

are satisfied.

Remark 3.1.1 *(i) We would like to mention that mass conservation of the Cahn-Larché system follows from the diffusion equation (3.9a) and the no-flux condition for $\mathbb{M}(Z)\nabla\mu$. Thus $\int_\Omega (c(t) - c^0)\,\mathrm{d}x = 0$ for all $t \in [0, T]$.*

(ii) To ensure that the solutions of (3.9)-(3.10) satisfy the constraint (3.3a), we have to impose

$$\sum_{k=1}^{N} \mathbb{M}_{kl}(Z) = 0 \ \textit{for all } l = 1, \ldots, N. \tag{3.11}$$

Indeed, this implies $\sum_{k=1}^{N}(\mathbb{M}(Z)\nabla\mu)_{kl} = 0$ for all $l = 1, \ldots, N$ and, therefore, the equation $\sum_{k=1}^{N}\left(\operatorname{div}(\mathbb{M}(Z)\nabla\mu)\right)_k = 0$.

(iii) By using the logarithmic potential (3.8) in the energy (3.4), the solutions of (3.9) - (3.10) will be forced to satisfy $c_k > 0$ for every component $k = 1, \ldots, N$. In particular, (3.3b) is fulfilled.

We see that property (3.11) implies $\mathbb{M}(Z) = \mathbb{M}(Z)\,\mathbb{P}$, where $\mathbb{P} \in \mathcal{L}(\mathbb{R}^N)$ denotes the orthogonal projection from $\mathbb{R}^N$ onto the subspace $\left\{x \in \mathbb{R}^N \mid x_1 + \ldots + x_N = 0\right\} \subseteq \mathbb{R}^N$. By setting $w := \mathbb{P}\mu$, the PDE (3.9a) can be rewritten as

$$\partial_t c = \operatorname{div}(\mathbb{M}(Z)\nabla\mu) = \operatorname{div}(\mathbb{M}(Z)\mathbb{P}\nabla\mu) = \operatorname{div}\left(\mathbb{M}(Z)\nabla(\mathbb{P}\mu)\right) = \operatorname{div}(\mathbb{M}(Z)\nabla w). \quad (3.12)$$

This formulation will be used for the analysis in Chapter 5 in the case of multi-component alloys.

Moreover, in order to guarantee existence of weak solutions, we will assume in Chapter 5 the *Onsager reciprocity law* which states that $\mathbb{M}(Z)$ has to be symmetric and that $\mathbb{M}(Z)$ is positive definite on the subspace $\left\{x \in \mathbb{R}^N \mid x_1 + \ldots + x_N = 0\right\}$, cf. [Gar00] (for constant mobility tensors).

3.2   Partial damage processes

Damage behavior in elastic materials originates from breaking atomic links which lead to the initiation and growing of microcracks and microvoids. Consequently, increasing damage lowers the integrity of the medium and, therefore, its material stiffness (see Figure 3.2 for a demonstration of crack growing after cyclic loading). A macroscopic phase-field model may specify the damage by a scalar quantity related to the density of such microdefects in the material. In the engineering literature, various macroscopic damage models have been extensively studied in the last decades, e.g. see [Fré02, Car86, Mie95, MK00, MS11, LD05, GUE+07], whereas mathematical works are more rarely and are in many cases based on the so-called *gradient-of-damage* model proposed by Frémond and Nedjar in [FN96]. The gradient term of the damage variable in the free energy has a smoothing effect from the mathematical point of view, i.e., it leads to smooth transitions between damaged and undamaged material states.

The gradient-of-damage model in [FN96] describes the damage progression by microscopic motions in the structures resulting from the growth of microcracks and microvoids. In contrast to a pure phenomenological modeling approach, the evolution laws for the damage processes are obtained from balance laws of the microscopic and the macroscopic forces together with some suitable constitutive relations. Based on the model developed in [FN96], existence and uniqueness results in the case of viscoelastic materials are proven in [BSS05] in the one dimensional case. Higher dimensional damage models are analytically investigated in [BS04, FK09, Gia05] for the parabolic and in [MR06, MT10, KRZ11] for the rate-independent case. We refer to [FG06, GL09, Bab11] for non-gradient approaches to damage models.

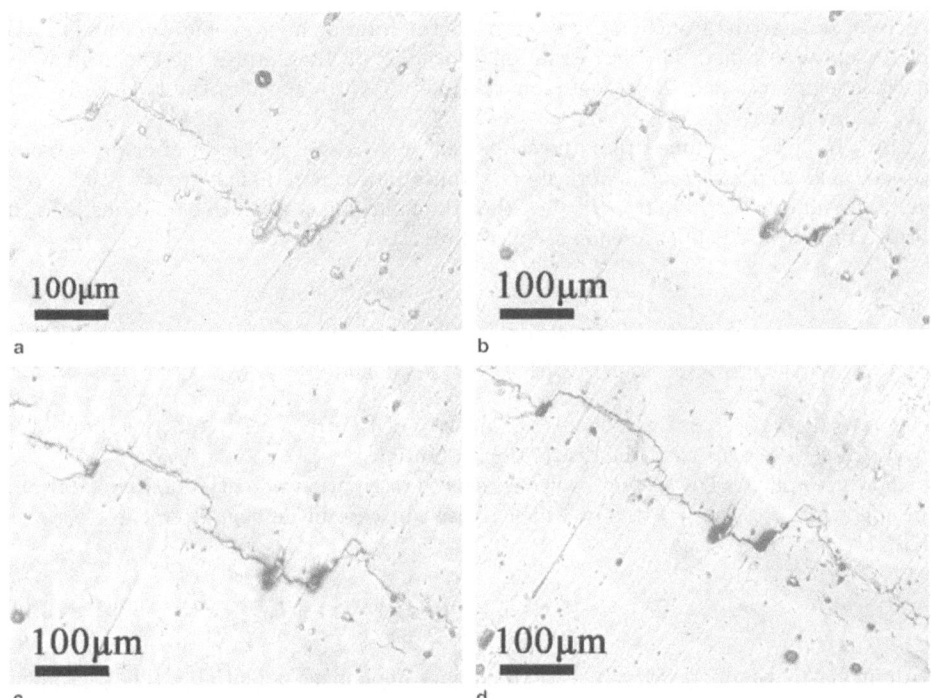

Figure 3.2: *Experiment with a laser sintered FeNiCu-alloy after four-point bending tests (see [WBB06]): (a) 23,000 cycles; (b) 25,000 cycles; (c) 29,000 cycles; (d) 31,000 cycles.*

The damage processes in this section and also in the most mathematical literature are assumed to be incomplete in the sense that even maximally damaged material parts still exhibit a minimal structural integrity, i.e., elastic properties. The complete damage case, in contrast, is treated in Section 3.3, and analytically in Chapter 6.

In the following, we summarize the approach in [FN96] to derive a PDE system for damage processes. The associated order parameter is denoted by z and should satisfy $0 \le z(x,t) \le 1$, where $z(x,t) = 1$ stands for a non-damaged and $z(x,t) = 0$ for a maximally damaged material point $(x,t) \in \Omega_T$. Furthermore, we make some modeling assumptions: The damage processes we want to consider in this paper are uni-directional, i.e., $\partial_t z(x,t) \le 0$. We assume a quasi-static force equilibrium, no external body and surface forces, and small strains, i.e., $e = \varepsilon(u) := \frac{1}{2}(\nabla u + (\nabla u)^{\mathrm{t}})$.

Then, the gradient-of-damage model is based on the following balance laws (see also [Fré12]):

$$\mathrm{div}\sigma = 0, \tag{3.13a}$$

$$\mathrm{div}H - B = 0, \tag{3.13b}$$

where σ denotes the Cauchy stress tensor, B the interior microscopic work and H the microscopic work flux. The first equation above, i.e. (3.13a), states the balance law for macroscopic forces and the second equation, i.e. (3.13b), specifies the balance law for microscopic forces.

In [FN96] it is assumed that the stress tensor σ as well as the interior microscopic work B and the microscopic work flux H splits into a non-dissipative σ^{nd}, B^{nd}, H^{nd} and a dissipative part σ^{d}, B^{d}, H^{d} such that the following constitutive relations hold (in terms of generalized subgradients; see also [Fré12])

$$\sigma = \sigma^{\mathrm{nd}} + \sigma^{\mathrm{d}} \text{ with } \sigma^{\mathrm{nd}} = \partial_e \psi \text{ and } \sigma^{\mathrm{d}} = \partial_{e_t} \phi, \tag{3.14a}$$

$$B = B^{\mathrm{nd}} + B^{\mathrm{d}} \text{ with } B^{\mathrm{nd}} \in \partial_z \psi \text{ and } B^{\mathrm{d}} \in \partial_{z_t} \phi, \tag{3.14b}$$

$$H = H^{\mathrm{nd}} + H^{\mathrm{d}} \text{ with } H^{\mathrm{nd}} \in \partial_{\nabla z} \psi \text{ and } H^{\mathrm{d}} \in \partial_{\nabla z_t} \phi, \tag{3.14c}$$

where $\psi = \psi(c, \nabla c, e, z, \nabla z)$ denotes the density of the free energy functional and $\phi = \phi(e_t, z_t, \nabla z_t)$ the density of the dissipation potential.

For an explicit PDE system, we consider the dissipation potential density (1.4) and the following free energy density which is used in certain damage literature (e.g. see [KRZ11]):

$$\psi(e, z, \nabla z) = \frac{1}{p} |\nabla z|^p + W^{\mathrm{el}}(e, z) + f(z) + I_{[0,\infty)}(z), \tag{3.15}$$

with an elastic energy density W^{el} and a damage dependent potential f. The p-gradient term in (3.15) accounts for microscopic interactions of the damage process, i.e., it has the consequence that a damaged material point influences its surroundings. From the mathematical point of view, the value of p determines the regularization of the system. We will investigate cases in which $p > n$ is assumed (resulting in Hölder continuous damage profiles) and cases in which p equals 2. The latter case is used in certain engineering literature [FN96, Fré12].

The elastic energy density for incomplete damage models with linear elasticity is given by

$$W^{\mathrm{el}}(e, z) = \frac{1}{2}(g(z) + \delta)e : \mathbf{C}e, \qquad \delta > 0, \tag{3.16}$$

where the function g with $g \geq 0$ describes the influence of the damage on the elastically stored energy. The positive constant δ ensures that the material can not completely disintegrate (partial damage). As in the previous section, the 4th order elasticity tensor $\mathbf{C}$ is assumed to be positive definite and the components satisfy the symmetry condition (3.5). Note that in the coupled system in Section 3.4, $\mathbf{C}$ will also depend on the concentration variable as in the previous section.

The contributions of the indicator functions $I_{[0,\infty)}(z)$ in the energy and $I_{(-\infty,0]}(z_t)$ in the dissipation potential lead to a doubly nonlinear differential inclusion formulation for (3.13b) which relates the derivative of the energy dissipation with the derivative of the free energy with respect to the damage variable.

Since the temperature is assumed to be constant and positive, the thermodynamic consistency is shown by validating the second law of thermodynamics. In the isothermal situation, this is equivalent to the following local dissipation inequality (also called Clausius-Duhem inequality; cf. [FN96]):

$$\partial_t \psi - \sigma : \partial_t e - B\partial_t z - H \cdot \partial_t \nabla z \leq 0. \tag{3.17}$$

Indeed, we will show that the dissipation inequality (3.17) is satisfied. Let us denote with $\widetilde{\psi}$ and $\widetilde{\phi}$ the free energy density (3.15) and the dissipation potential (1.4) without their indicator functions. Furthermore, let us denote, for the moment, with ∂_t the time derivative from the *left side* (see [FN96], where the time derivative of z might be discontinuous). Applying the chain rule yields (set $Z := (e, z, \nabla z)$)

$$\partial_t \psi(Z) = \partial_e \psi(Z) : \partial_t e + \partial_z \widetilde{\psi}(Z)\partial_t z + \partial_{\nabla z}\psi(Z) \cdot \partial_t \nabla z. \tag{3.18}$$

Beyond that, we find the inequalities

$$\zeta \cdot \partial_t z \geq 0, \tag{3.19a}$$

$$(\partial \widetilde{\phi}(\partial_t z) + \varrho) \cdot \partial_t z \geq 0 \tag{3.19b}$$

for every $\zeta \in \partial I_{[0,\infty)}(z)$ and $\varrho \in I_{(-\infty,0]}(\partial_t z)$. Adding (3.18), (3.19a) and (3.19b), we obtain the inequality

$$\partial_t \psi(Z) - \partial_e \psi(Z) : \partial_t e - (\partial_z \widetilde{\psi}(Z) + \zeta + \partial \widetilde{\phi}(\partial_t z) + \varrho)\partial_t z - \partial_{\nabla z}\psi(Z) \cdot \partial_t \nabla z \leq 0.$$

By setting

$$\sigma^{\mathrm{nd}} := \partial_e \psi(Z), \qquad B^{\mathrm{nd}} := \partial_z \widetilde{\psi}(Z) + \zeta \in \partial_z \psi(Z), \qquad H^{\mathrm{nd}} := \partial_{\nabla z}\psi(Z),$$
$$\sigma^{\mathrm{d}} := \partial_{e_t}\phi(\partial_t z) = 0, \qquad B^{\mathrm{d}} := \partial\widetilde{\phi}(\partial_t z) + \varrho \in \partial_{z_t}\phi(\partial_t z), \qquad H^{\mathrm{d}} := \partial_{\nabla z_t}\phi(\partial_t z) = 0,$$

the dissipation inequality (3.17) is proven. This shows that the gradient-of-damage model above is thermodynamically consistent.

At the end of this section, we will give a classical formulation of the partial damage problem consisting of the balance laws (3.13), the constitutive relations (3.14), the free energy (3.15) and the dissipation potential density (1.4) supplemented with suitable initial-boundary conditions.

We call a pair of functions (u, z) with $u : \overline{\Omega_T} \to \mathbb{R}^n$ and $z : \overline{\Omega_T} \to \mathbb{R}$ a classical solution to our partial damage problem with the initial-boundary data (z^0, b) if

$$0 = \mathrm{div}\big(\partial_e W^{\mathrm{el}}(\epsilon(u), z)\big) \qquad\qquad \text{in } \Omega_T, \tag{3.20a}$$

$$0 = -\Delta_p z + \partial_z W^{\mathrm{el}}(\epsilon(u), z) + f'(z) + \zeta - \alpha + \beta\partial_t z + \varrho \qquad \text{in } \Omega_T \tag{3.20b}$$

with $\zeta \in \partial I_{[0,\infty)}(z)$ and $\varrho \in \partial I_{(-\infty,0]}(\partial_t z)$ and the initial-boundary conditions

$$z(0) = z^0 \qquad\qquad \text{in } \Omega, \qquad (3.21\text{a})$$

$$u = b \qquad\qquad \text{on } (\Gamma_\text{D})_T, \qquad (3.21\text{b})$$

$$\partial_e W^\text{el} \cdot \nu = 0 \qquad\qquad \text{on } (\Gamma_\text{N})_T, \qquad (3.21\text{c})$$

$$\nabla z \cdot \nu = 0 \qquad\qquad \text{on } (\partial\Omega)_T \qquad (3.21\text{d})$$

are satisfied.

Remark 3.2.1   *(i) This PDE system is called rate-dependent because of the quadratic contribution of z_t in the dissipation potential ϕ in (1.4) which prevents ϕ from being positively 1-homogeneous. Otherwise, in the case $\beta = 0$, we call the system rate-independent, see [KRZ11] for more details. We would also like to refer to [Mie05, EM06, MRZ10, MT10] for analytical investigations of rate-independent damage system and numerical simulations.*

(ii) Due to the subgradients of the indicator functions in (3.20b), we obtain the restrictions $0 \leq z \leq 1$ and $\partial_t z \leq 0$ provided that the initial value z^0 satisfies $0 \leq z^0 \leq 1$.

3.3   Complete damage processes

In complete damage models, the elastic material is allowed to completely disintegrate. Roughly speaking, in this regime, the evolution is governed by the same PDE system (3.20) but with a degenerating elastic energy density (3.16) (with $\delta = 0$). In most of the mathematical damage literature, one usually assumes partial damage behavior and dropping this assumption gives rise to various mathematical challenges. Mathematical works of such models covering global-in-time existence are rarely and are mainly focused on purely *rate-independent systems* [MR06, BMR09, MRZ10, Mie11] by using Γ-convergence techniques to recover energetic properties in the limit. Modeling aspects and existence of weak solutions for purely mechanical complete damage systems with quasi-static force balances are studied in [BMR09] and with viscoelasticity in [MRZ10, RR12]. Existence results for *rate-dependent* complete damage systems in thermoviscoelastic materials are shown in [RR12].

The reason why incomplete damage models are more feasible for mathematical investigations is that a uniform convexity assumption on the elastic energy density (see (4.2a)) prevents the material from a complete degeneration. By rejecting this assumption, the deformation variable would become meaningless on material fragments with maximal damage because the elastic energy density vanishes regardless of the values of u (see for instance [BMR09]). However, in the case of viscoelastic materials, the inertia terms in the momentum balance equation circumvent this kind of problem in the sense that the displacement field still exists on the whole domain accompanied with a loss of spatial regularity (cf. [MRZ10, RR12]).

In the following, we are going to explain our proposed model for rate-dependent complete damage processes where the forces are assumed to be in a quasi-static equilibrium. To the author's best knowledge, this case is not covered in the mathematical literature so far. The free energy and the dissipation potential density are of the same type as in (3.15) and (1.4), respectively. The elastic energy density is assumed to be of the form

$$W^{\text{el}}(e, z) = \frac{1}{2}g(z)e : \mathbf{C}e \tag{3.22}$$

with a differentiable function g satisfying $g \geq 0$ and $g' \geq \eta$ for a constant $\eta > 0$ and a positive definite 4th order stiffness tensor $\mathbf{C}$ satisfying (3.5). Note that complete damage is possible if and only if $g(0) = 0$. On the contrary, the case $g(0) > 0$ describes incomplete damage processes which are treated in the previous section (and analytically in Chapter 4 and Chapter 5). Due to the degenerated elastic energy density (3.22), the momentum and the microforce balance laws (3.20) are only meaningful in the area where $z > 0$. Beyond that, as already mentioned in the introduction, a phenomenon (referred to as *material exclusion* in the following) might cause severe mathematical difficulties. The reason is explained below.

Suppose that (u_δ, z_δ) is a solution (in a certain weak sense - see Chapter 4) of (3.20)-(3.21) for $\delta > 0$ in the elastic energy density (3.16). Via certain a-priori estimates and compactness theorems, we can identify a clusterpoint z such that $z_\delta \to z$ in some weak sense as $\delta \to 0^+$ for a subsequence. Suppose now that at a specific time point t, a path-connected component P from $\{x \in \overline{\Omega} \,|\, z(x, t) > 0\}$ is isolated from the Dirichlet boundary, i.e., $\mathcal{H}^{n-1}(P \cap \Gamma_{\mathrm{D}}) = 0$. In this case, Korn's inequality in Theorem 2.3.8 is not available on P (even if P is a smooth domain) and, consequently, the displacement field u_δ for the regularized system cannot be controlled on P in the degenerate limit $\delta \to 0^+$. To overcome this problem, path-connected components P of the not completely damaged area $\{z(t) > 0\}$ isolated from the Dirichlet boundary, i.e. $\mathcal{H}^{n-1}(P \cap \Gamma_{\mathrm{D}}) = 0$, will be excluded from the domain. On the one hand, this make our model accessible for a rigorous analysis and, on the other hand, for some applications,

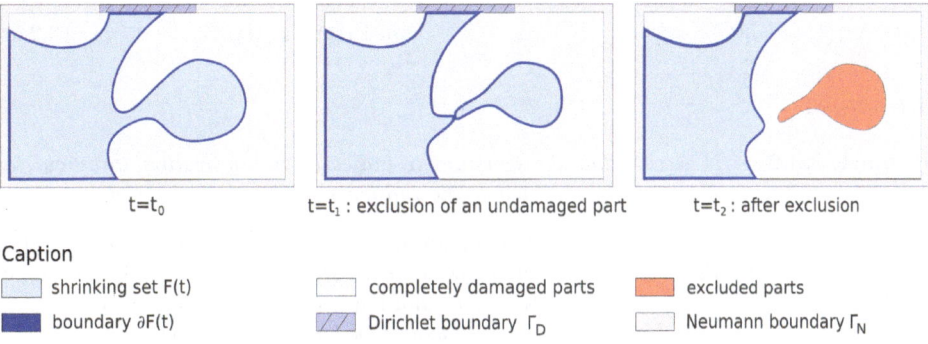

| | $t=t_0$ | | $t=t_1$: exclusion of an undamaged part | | $t=t_2$: after exclusion |

Caption

| | shrinking set F(t) | | completely damaged parts | | excluded parts |
| | boundary ∂F(t) | | Dirichlet boundary Γ_{D} | | Neumann boundary Γ_{N} |

Figure 3.3: *This illustration shows the exclusion of an undamaged material part in 2D during the evolution process. The dark blue curve encircles the maximal admissible subset $\mathfrak{A}_{\Gamma_{\mathrm{D}}}(\{z(t) > 0\})$ of the not completely damaged area $F(t) := \{z(t) > 0\}$ (see Definition 2.4.3).*

the detached parts might be of little interest anyway. This approach is illustrated in Figure 3.3 and motivates the definition of maximal admissible subsets with respect to Γ_D which has already been given in Definition 2.4.3.

The admissibility condition for the time-dependent domain is displayed in Figure 3.3 and can be formalized in the following way:

Due to the monotonicity of the damage function z with respect to time, the set $F \subseteq \overline{\Omega_T}$ given by $F(t) := \{z(t) > 0\}$ is shrinking (see Section 2.4).

In a nutshell, the evolutionary problem in the last section will be considered on a time-dependent and shrinking domain (see Section 2.4) which coincides with the not completely damaged area and which is, for any time, admissible with respect to Γ_D. The associated PDE system with its time-dependent domain and its initial-boundary conditions can be summarized within a classical notion in the following way.

Definition 3.3.1 (Classical solution for complete damage) *A pair of functions (u, z) with $u \in C_x^2(F; \mathbb{R}^n)$, $z \in C^2(\overline{\Omega_T}; \mathbb{R})$, where the shrinking set F is given by $F := \mathfrak{A}_{\Gamma_D}(\{z > 0\})$ is called a classical solution to the initial-boundary data (z^0, b) if*

$$0 = \mathrm{div}(\partial_e W^{\mathrm{el}}(\epsilon(u), z)) \qquad\qquad\qquad\qquad\qquad \text{in } F, \qquad (3.23\mathrm{a})$$

$$0 = -\mathrm{div}(|\nabla z|^{p-2}\nabla z) + \partial_z W^{\mathrm{el}}(\epsilon(u), z) + f'(z) - \alpha + \beta\partial_t z + \zeta \quad \text{in } F, \qquad (3.23\mathrm{b})$$

with $\zeta \in \partial I_{(-\infty,0]}(\partial_t z)$ and the initial-boundary conditions

$$z(0) = z^0 \qquad\qquad\qquad\qquad \text{in } F(0), \qquad\qquad\qquad\qquad\qquad\quad (3.24\mathrm{a})$$

$$u(t) = b(t) \qquad\qquad\qquad\qquad \text{on } \Gamma_1(t) := F(t) \cap \Gamma_D, \qquad\qquad (3.24\mathrm{b})$$

$$\partial_e W^{\mathrm{el}}(\epsilon(u(t)), z(t)) \cdot \nu = 0 \quad\;\; \text{on } \Gamma_2(t) := F(t) \cap \Gamma_N, \qquad\qquad (3.24\mathrm{c})$$

$$z(t) = 0 \qquad\qquad\qquad\qquad\quad \text{on } \Gamma_3(t) := \partial F(t) \setminus F(t), \qquad\quad (3.24\mathrm{d})$$

$$\nabla z(t) \cdot \nu = 0 \qquad\qquad\qquad \text{on } \Gamma_1(t) \cup \Gamma_2(t), \qquad\qquad\qquad\;\; (3.24\mathrm{e})$$

are satisfied.

Remark 3.3.2 *(i) Recall that the constant α indicates the activation threshold for the damage process whereas β models the influence of the viscous effects.*

(ii) The time-dependent boundary $\partial F(t)$ disjointly decomposes into $\Gamma_1(t) \cup \Gamma_2(t) \cup \Gamma_3(t)$, where $\Gamma_1(t)$ indicates the not completely damaged Dirichlet boundary, $\Gamma_2(t)$ the not completely damaged Neumann boundary and $\Gamma_3(t)$ the completely damaged boundary (see Figure 3.4). We have the following types of boundary conditions:

- $\Gamma_1(t)$ —   *Dirichlet boundary condition for u,*
        Neumann boundary condition for z,

- $\Gamma_2(t)$ —   *Neumann boundary condition for u,*
        Neumann boundary condition for z,

- $\Gamma_3(t)$ —   *degenerated boundary condition.*

On the degenerated boundary, z vanishes (homogeneous Dirichlet boundary condition for z) and, therefore, if we assume that $\epsilon(u)$ can be continuously extended to Γ_3 the stress $\partial_\epsilon W^{el}(\epsilon(u), z)$ vanishes too.

(iii) *Since $z > 0$ in F, the subdifferential $\partial I_{[0,\infty)}(z)$ vanishes in the differential inclusion (3.23b) in comparison to (3.20b).*

(iv) *The regularity assumption for z in this definition is very generous. For instance, it suffices to use left time derivatives for z in the PDE system in order to allow for jumps in the time derivative of z (see [FN96]).*

The thermodynamic consistency follows with the same argumentation as in Section 3.2. Existence results for the system in Definition 3.3.1 are provided in Chapter 6.

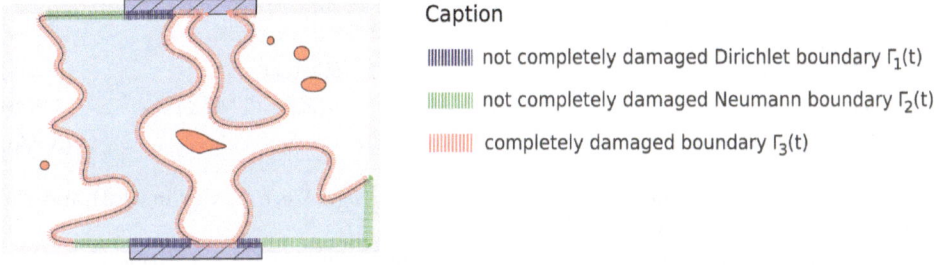

Caption

▨ not completely damaged Dirichlet boundary $\Gamma_1(t)$

▨ not completely damaged Neumann boundary $\Gamma_2(t)$

▨ completely damaged boundary $\Gamma_3(t)$

Figure 3.4: *Illustration of the different parts of the boundary of $F(t)$.*

3.4 Phase separation coupled with damage processes

Much of the recent literature for phase separation is focused on the couplings of Cahn-Hilliard and of Allen-Cahn type models to broaden the possible range of application areas. The analysis and numerics of the coupled system often becomes much more involved due to new dependencies and variables entering the equations. For instance, Cahn-Hilliard equations have been coupled with temperature equations in [MS05, AP92, GR07], with Navier-Stokes equations in [BB11] and with Allen-Cahn equations in [BdS04, BGN07]. In the previous sections, we have reviewed phase-field models for phase separation and damage processes separately. This section is dedicated to couple these systems in one unifying model to account for both phenomena. As mentioned in Chapter 1, experimental data reveal that phase separation may lead to critical stresses at phase boundaries which, in turn, result in crack and void formations. This phenomenon is of particular interest to understand the aging process in solder materials.

Here, we qualify our model formally and postpone a rigorous treatment to Chapter 4, Chapter 5 and Chapter 7. For the classical formulation of the coupled system, we will combine the approaches presented in Section 3.1 and Section 3.2. The resulting model is based on the balance laws (3.1) for phase separation and the balance laws (3.13) for the damage processes in combination with suitable constitutive relations. More specifically, the considered balance laws are given by

$$\partial_t c = -\operatorname{div} h, \tag{3.25a}$$

$$\operatorname{div} \xi + \pi = 0, \tag{3.25b}$$

$$\operatorname{div} \sigma = 0, \tag{3.25c}$$

$$\operatorname{div} H - B = 0. \tag{3.25d}$$

Equations (3.25a) and (3.25b) are mass and microforce balance laws for the Cahn-Hilliard equations (see Section 3.1), (3.25c) is the momentum balance law for a quasi-static equilibrium, and (3.25d) is the microforce balance equation for the damage propagation law (see Section 3.2). We impose the following constitutive relations, cf. (3.2) and (3.14) (set $Z := (c, \nabla c, e, z, \nabla z, \mu, \nabla \mu)$):

$$\partial_c \psi(Z) = \mu - \pi, \qquad \partial_{\nabla c} \psi(Z) = \xi, \qquad \partial_e \psi(Z) = \sigma, \tag{3.26a}$$

$$\partial_\mu \psi(Z) = 0, \qquad \partial_{\nabla \mu} \psi(Z) = 0, \qquad h(Z) = -\mathbb{M}(Z) \nabla \mu, \tag{3.26b}$$

$$\partial_z \psi(Z) + \partial_{z_t} \phi(\partial_t z) \ni B, \qquad \partial_{\nabla z} \psi(Z) = H, \tag{3.26c}$$

as well as the unifying energy density function $\psi = \psi(c, \nabla c, e, z, \nabla z)$ in (1.3) and the damage dissipation potential density $\phi = \phi(z_t)$ in (1.4).

The thermodynamic consistency is shown via the chain rule, the constitutive relations (3.26), and by taking into account the inequalities (3.19a) and (3.19b) ($\zeta \in \partial I_{[0,\infty)}(z)$ and $\varrho \in I_{(-\infty,0]}(\partial_t z)$ are arbitrary for the moment). Let $\widetilde{\psi}$ be the free energy density (1.3) and $\widetilde{\phi}$ the dissipation potential density (1.4) without their indicator functions. We obtain

$$\partial_t \psi(Z) = \partial_c \psi(Z) \cdot \partial_t c + \partial_{\nabla c} \psi(Z) : \partial_t \nabla c + \partial_e \psi(Z) : \partial_t e + \partial_z \widetilde{\psi}(Z) \partial_t z + \partial_{\nabla z} \psi(Z) \cdot \partial_t \nabla z$$
$$\leq (\mu - \pi) \partial_t c + \xi : \partial_t \nabla c + \sigma : \partial_t e + (\partial_z \widetilde{\psi}(Z) + \zeta + \partial \widetilde{\phi}(\partial_t z) + \varrho) \partial_t z$$
$$+ H \cdot \partial_t \nabla z - h : \nabla \mu.$$

Choosing ζ and ϱ such that $B = \partial_z \widetilde{\psi}(Z) + \zeta + \partial \widetilde{\phi}(\partial_t z) + \varrho$ (see (3.26)), we end up with the desired dissipation inequality

$$\partial_t \psi(Z) - \sigma : \partial_t e + (\pi - \mu) \partial_t c - \xi : \partial_t \nabla c - B \partial_t z - H \cdot \partial_t \nabla z + h : \nabla \mu \leq 0.$$

Plugging in the density functions ψ and ϕ given in (1.3) and (1.4) into the constitutive relations (3.26), the balance laws (3.25) become the following elliptic-parabolic system of differential equations and differential inclusion:

Definition 3.4.1 (Classical solution for coupled system with partial damage)
A tuple of functions (c, u, z, w) with $c \in \mathcal{C}^2(\overline{\Omega_T}; \mathbb{R}^N)$, $u \in \mathcal{C}^2_x(\overline{\Omega_T}; \mathbb{R}^n)$, $z \in \mathcal{C}^2(\overline{\Omega_T}; \mathbb{R})$

and $w \in \mathcal{C}^2(\overline{\Omega_T}; \mathbb{R}^N)$ is called a classical solution to the initial-boundary data (c^0, z^0, b) if

$$\partial_t c = \operatorname{div}(\mathbb{M}(Z)\nabla w) \ with \ w = \mathbb{P}\mu \qquad \qquad in \ \Omega_T, \qquad (3.27\text{a})$$

$$\mu = -\operatorname{div}(\Gamma \nabla c) + \partial_c W^{\text{ch}}(c) + \partial_c W^{\text{el}}(c, \epsilon(u), z) \qquad in \ \Omega_T, \qquad (3.27\text{b})$$

$$\operatorname{div}\big(\partial_e W^{\text{el}}(c, \epsilon(u), z)\big) = 0 \qquad \qquad in \ \Omega_T, \qquad (3.27\text{c})$$

$$0 = -\Delta_p z + \partial_z W^{\text{el}}(c, \epsilon(u), z) + f'(z) + \zeta - \alpha + \beta \partial_t z + \varrho \quad in \ \Omega_T \qquad (3.27\text{d})$$

with $\zeta \in \partial I_{[0,\infty)}(z)$ and $\varrho \in \partial I_{(-\infty,0]}(\partial_t z)$ and the initial-boundary conditions

$$c(0) = c^0 \qquad \qquad in \ \Omega, \qquad (3.28\text{a})$$

$$z(0) = z^0 \qquad \qquad in \ \Omega, \qquad (3.28\text{b})$$

$$u = b \qquad \qquad on \ (\Gamma_{\mathrm{D}})_T, \qquad (3.28\text{c})$$

$$\partial_e W^{\text{el}} \cdot \nu = 0 \qquad \qquad on \ (\Gamma_{\mathrm{N}})_T, \qquad (3.28\text{d})$$

$$\mathbb{M}(Z)\nabla w \cdot \nu = 0 \qquad \qquad on \ (\partial \Omega)_T, \qquad (3.28\text{e})$$

$$\Gamma \nabla c \cdot \nu = 0 \qquad \qquad on \ (\partial \Omega)_T, \qquad (3.28\text{f})$$

$$\nabla z \cdot \nu = 0 \qquad \qquad on \ (\partial \Omega)_T \qquad (3.28\text{g})$$

are satisfied.

Remark 3.4.2 *As in Section 3.1, we have used the formulation (3.12), where the matrix $\mathbb{P}$ denotes the orthogonal projection of $\mathbb{R}^N$ onto the subspace $\{x \in \mathbb{R}^N \mid \sum_{k=1}^{N} x_k = 0\}$.*

In a nutshell, equations (3.27a)-(3.27b) are fourth order parabolic equations of Cahn-Hilliard type and describe phase separation processes for the concentration c while the elliptic equation (3.27c) constitutes a quasi-static mechanical equilibrium for u. The doubly nonlinear differential inclusion (3.27d) specifies the flow rule for the damage profile according to the constraints $z \geq 0$ and $\partial_t z \leq 0$, see Section 3.2. In Chapter 4 and in Chapter 5, we will prove existence of weak solutions for the system given in Definition 3.4.1.

Let us remark that the above system is only a suitable formulation for partial damage processes (see discussion in Section 3.3). We conclude this section by considering the complete damage case, i.e., (1.5) with $\delta = 0$. To this end, we adopt the ideas from Section 3.3 concerning the notion of maximal admissible subsets and express our system with the help of a time-dependent domain.

Definition 3.4.3 (Classical solution for coupled system with complete damage)
A tuple of functions (c, u, z, w) with $c \in \mathcal{C}^2(\overline{F}; \mathbb{R})$, $u \in \mathcal{C}_x^2(F; \mathbb{R}^n)$, $z \in \mathcal{C}^2(\overline{\Omega_T}; \mathbb{R})$ and $w \in \mathcal{C}^2(F; \mathbb{R}^N)$ and the shrinking set $F \subseteq \overline{\Omega_T}$ with $F := \mathfrak{A}_{\Gamma_{\mathrm{D}}}(\{z > 0\})$ is called a classical solution to the initial-boundary data (c^0, z^0, b) if

$$\partial_t c = \text{div}(\mathbb{M}(Z)\nabla w) \ \textit{with } w = \mathbb{P}\mu \qquad\qquad\qquad in\ F, \qquad (3.29a)$$
$$\mu = -\text{div}(\Gamma\nabla c) + \partial_c\Psi(c) + \partial_c W^{\text{el}}(c, \epsilon(u), z) \qquad\qquad in\ F, \qquad (3.29b)$$
$$0 = \text{div}(\partial_e W^{\text{el}}(c, \epsilon(u), z)) \qquad\qquad\qquad in\ F, \qquad (3.29c)$$
$$0 = -\text{div}(|\nabla z|^{p-2}\nabla z) + \partial_z W^{\text{el}}(c, \epsilon(u), z) + f'(z) - \alpha + \beta\partial_t z + \zeta \ in\ F \quad (3.29d)$$

with $\zeta \in \partial I_{(-\infty,0]}(\partial_t z)$ and the initial-boundary conditions

$$c(0) = c^0 \qquad\qquad\qquad\qquad in\ F(0), \qquad\qquad\quad (3.30a)$$
$$z(0) = z^0 \qquad\qquad\qquad\qquad in\ F(0), \qquad\qquad\quad (3.30b)$$
$$u(t) = b(t) \qquad\qquad\qquad\qquad on\ \Gamma_1(t) := F(t) \cap \Gamma_{\text{D}}, \quad (3.30c)$$
$$\partial_e W^{\text{el}}(c(t), \epsilon(u(t)), z(t)) \cdot \nu = 0 \qquad on\ \Gamma_2(t) := F(t) \cap \Gamma_{\text{N}}, \quad (3.30d)$$
$$z(t) = 0 \qquad\qquad\qquad\qquad on\ \Gamma_3(t) := \partial F(t) \setminus F(t), \quad (3.30e)$$
$$\nabla z(t) \cdot \nu = 0 \qquad\qquad\qquad on\ \Gamma_1(t) \cup \Gamma_2(t), \qquad (3.30f)$$
$$\Gamma\nabla c(t) \cdot \nu = 0 \qquad\qquad\qquad on\ \Gamma_1(t) \cup \Gamma_2(t), \qquad (3.30g)$$
$$\mathbb{M}(Z(t))\nabla w(t) \cdot \nu = 0 \qquad\qquad on\ \Gamma_1(t) \cup \Gamma_2(t) \qquad (3.30h)$$

are satisfied.

Existence results for this system are shown in Chapter 7. There, we will impose the degeneracy condition (7.5). This prevents the phase separation process from diffusing chemical substances through completely damaged material.

CHAPTER 4

Cahn-Hilliard systems with polynomial chemical potentials coupled with damage processes and homogeneous elasticity

The present chapter covers certain existence results for Cahn-Hilliard equations which are coupled with elasticity and partial damage processes (see Definition 3.4.1). We assume a binary mixture, a polynomial growth condition for the chemical potential, a homogeneous elastic energy density (with respect to the chemical concentration) and a p-Laplacian with $p > n$ in the differential inclusion for the damage propagation law.

The aim of the chapter is twofold: Firstly, we develop a notion of weak solution in Section 4.2 which requires less regularity of the solutions than the subdifferential formulation. The formulation consists of variational properties and an energy estimate.

Secondly, in the main part, Section 4.3, existence of weak solutions are proven. On the one hand, several well-established methods such as semi-implicit time-discretization, direct methods in the calculus of variations, energy a-priori estimates and compactness results from Aubin and Lions are used. But on the other hand, some new methods which are introduced in this work such as test function approximations (see Lemma 2.3.18), characterization of variational inequalities (see Lemma 2.3.19) and certain regularization techniques are utilized to handle the variational inequalities and the energy inequality.

The results and proofs in this chapter are published in [HK11].

4.1   Assumptions

To study the coupled PDE system (3.27) with the initial-boundary conditions (3.28), we have to make some assumptions.

We assume a binary mixture. In this case, c reduces to a real valued function since it suffices to describe the concentration of one of the two components and the Cahn-Hilliard equations (3.27a)-(3.27b) read as

$$\partial_t c = \operatorname{div}(\mathbb{M}(Z)\nabla\mu),$$
$$\mu = -\operatorname{div}(\Gamma\nabla c) + \partial_c W^{\mathrm{ch}}(c) + \partial_c W^{\mathrm{el}}(c, \epsilon(u), z).$$

Furthermore (see Section 3.4 for an explanation of the functions):

(i) *Energy density functions and gradient tensors.*

 We assume

$$W^{\mathrm{el}} \in \mathcal{C}^1(\mathbb{R} \times \mathbb{R}^{n \times n} \times \mathbb{R}; \mathbb{R}_+), \tag{4.1a}$$
$$W^{\mathrm{ch}} \in \mathcal{C}^1(\mathbb{R}; \mathbb{R}_+), \tag{4.1b}$$
$$\mathbb{M} = \Gamma = \mathrm{Id}, \tag{4.1c}$$
$$f = 0, \tag{4.1d}$$
$$p > n. \tag{4.1e}$$

(ii) *Convexity, structural and growth assumptions.*

 The functions W^{el} and W^{ch} are assumed to satisfy for some constants $\eta > 0$ and $C > 0$ the following estimates:
 for arbitrary $c \in \mathbb{R}$, $z \in \mathbb{R}$ and symmetric $e, e_1, e_2 \in \mathbb{R}^{n \times n}$:

$$\eta|e_1 - e_2|^2 \leq (\partial_e W^{\mathrm{el}}(c, e_1, z) - \partial_e W^{\mathrm{el}}(c, e_2, z)) : (e_1 - e_2), \tag{4.2a}$$
$$W^{\mathrm{el}}(c, e, z) = W^{\mathrm{el}}(c, e^t, z), \tag{4.2b}$$
$$W^{\mathrm{el}}(c, e, z) \leq C(|c|^2 + |e|^2 + 1), \tag{4.2c}$$
$$|\partial_c W^{\mathrm{el}}(c, e, z)| \leq C(|c|^2 + |e| + 1), \tag{4.2d}$$
$$|\partial_e W^{\mathrm{el}}(c, e_1 + e_2, z)| \leq C(W^{\mathrm{el}}(c, e_1, z) + |e_2| + 1), \tag{4.2e}$$
$$|\partial_z W^{\mathrm{el}}(c, e, z)| \leq C(|c|^2 + |e|^2 + 1), \tag{4.2f}$$
$$|\partial_c W^{\mathrm{ch}}(c)| \leq C(|c|^{2^\star/2} + 1). \tag{4.2g}$$

For dimension $n = 3$, the constant $2^\star$ denotes the Sobolev critical exponent. In the two dimensional case, the constant $2^\star$ can be an arbitrary positive real number and in one space dimension (4.2g) can be dropped.

Remark 4.1.1   *(i) Condition (4.2e) with $e_1 = 0$ and $e_2 = e$ as well as condition (4.2c) imply the estimate*

$$|\partial_e W^{\mathrm{el}}(c, e, z)| \leq C(W^{\mathrm{el}}(c, 0, z) + |e| + 1) \leq C(|c|^2 + |e| + 1). \tag{4.3}$$

A lower bound for $W^{el}(c, e, z)$ can be obtained as in [Gar00, Section 3.2]. To this end, we write

$$W^{el}(c, e, z) = W^{el}(c, 0, z) + \int_0^1 \partial_e W(c, te, z) : e \, dt$$

and use (4.2a) with $e_1 = te$ and $e_2 = 0$, i.e.,

$$\partial_e W^{el}(c, te, z) : e \geq t\eta |e|^2 + \partial_e W^{el}(c, 0, z) : e.$$

Combining both calculations, using (4.2c) and (4.3) as well as Young's inequality, we obtain ($\eta_1, \eta_2, \eta_3 > 0$, denote positive constants independent of c, u and z)

$$
\begin{aligned}
W^{el}(c, e, z) &= W^{el}(c, 0, z) + \int_0^1 \partial_e W(c, te, z) : e \, dt \\
&\geq W(c, 0, z) + \frac{1}{2}\eta |e|^2 + \partial_e W^{el}(c, 0, z) : e \\
&\geq -\eta_1(|c|^2 + 1) + \frac{1}{2}\eta |e|^2 - \frac{1}{4}\eta |e|^2 - \frac{1}{\eta}|\partial_e W^{el}(c, 0, z)|^2 \\
&\geq -\eta_1(|c|^2 + 1) + \frac{1}{4}\eta |e|^2 - \eta_2(|c|^2 + 1)^2 \\
&= \frac{1}{4}\eta |e|^2 - \eta_3(|c|^4 + 1).
\end{aligned}
\tag{4.4}
$$

(ii) *It is important to note that homogeneous elastic energy densities of the type*

$$W^{el}(c, e, z) = \frac{1}{2}(g(z) + \delta)\mathbf{C}(e - e^\star(c)) : (e - e^\star(c)),$$

where Vegard's law is assumed (consequently, $|\partial_c e^\star(c)|$ is bounded; see (3.6)) and $g \in \mathcal{C}^1(\mathbb{R}; \mathbb{R}_+)$ and $g' \in \mathcal{C}(\mathbb{R})$ are assumed to be bounded, satisfy the assumptions above. More precisely, for all $c \in \mathbb{R}$, $z \in \mathbb{R}$ and symmetric $e \in \mathbb{R}^{n \times n}$:

$$
\begin{aligned}
|\partial_c W^{el}(c, e, z)| &= |(g(z) + \delta)\mathbf{C}(e - e^\star(c)) : \partial_c e^\star(c)| \leq C|(g(z) + \delta)\mathbf{C}(e - e^\star(c))| \\
&\leq C(|c| + |e|), \\
|\partial_e W^{el}(c, e_1 + e_2, z)| &= |(g(z) + \delta)\mathbf{C}(e_1 - e^\star(c)) + (g(z) + \delta)\mathbf{C}e_2|, \\
&\leq C(1 + W^{el}(c, e_1, z) + |e_2|), \\
|\partial_z W^{el}(c, e, z)| &= \left|\frac{1}{2}g'(z)\mathbf{C}(e - e^\star(c)) : (e - e^\star(c))\right| \leq C(|c|^2 + |e|^2).
\end{aligned}
$$

Concentration dependent stiffness tensors $\mathbf{C}(c)$, on the other hand, are treated in Chapter 5.

(iii) *Without loss of generality, the damage potential density function f in (1.3) is assumed to be 0. But the existence proof in Section 4.3 also works for every $f \in \mathcal{C}^1(\mathbb{R}; \mathbb{R}_+)$ function if f and f' are bounded. More precisely, the potential f can* *also be incorporated in the elastic energy density function W^{el} without violating the assumptions (4.2). Indeed, consider the energy*

$$\widetilde{W}^{\text{el}}(c, e, z) := W^{\text{el}}(c, e, z) + f(z).$$

Since f and f' are bounded, the function $\widetilde{W}^{\text{el}}(c, e, z)$ satisfies the growth assumptions (4.2) if $W^{\text{el}}(c, e, z)$ satisfies the growth assumptions (4.2).

(iv) *For $n = 3$, it follows $2^\star = 6$. Due to (4.2g), polynomial functions of degree 3 are allowed for $\partial_c W^{\text{ch}}$ and, consequently, polynomial functions of degree 4 can be used for W^{ch}. Therefore, double well potential functions of type (3.7) can be employed.*

4.2   Weak formulations and existence results

The analytical treatment of the system in Definition 3.4.1 with the assumptions (4.1) and (4.2) in this chapter uses a regularization method that gives a better regularity property for c and a better integrability for u in the first instance. A passage to the limit will finally give us weak solutions to the original problem.

In doing so, we develop in Subsection 4.2.1 and Subsection 4.2.2 a notion of weak solutions which consists of variational equalities and inequalities as well as an energy estimate. This notion was inspired by the concept of energetic solutions in the framework of rate-independent systems. The advantage is that solutions in this weak formulation can be obtained by certain variational methods and lower semi-continuity arguments.

For the analysis, it is convenient to introduce the energy functional

$$\mathcal{E} : H^1(\Omega) \times H^1(\Omega; \mathbb{R}^n) \times W^{1,p}(\Omega) \to \mathbb{R}_\infty$$

and the dissipation potential

$$\mathcal{R} : L^2(\Omega) \to \mathbb{R}_\infty$$

defined by their densities (1.3) and (1.4) as

$$\mathcal{E}(c, u, z) := \int_\Omega \psi(c, \nabla c, \epsilon(u), z, \nabla z) \, \mathrm{d}x$$

$$= \int_\Omega \left(\frac{1}{2} |\nabla c|^2 + \frac{1}{p} |\nabla z|^p + W^{\text{el}}(c, \epsilon(u), z) + W^{\text{ch}}(c) + I_{[0,\infty)}(z) \right) \mathrm{d}x, \quad (4.5a)$$

$$\mathcal{R}(z_t) := \int_\Omega \phi(z_t) \, \mathrm{d}x = \int_\Omega \left(-\alpha z_t + \frac{\beta}{2} |z_t|^2 + I_{(-\infty,0]}(z_t) \right) \mathrm{d}x \quad (4.5b)$$

as well as

$$\widetilde{\mathcal{E}}(c, u, z) := \int_\Omega \left(\frac{1}{2} |\nabla c|^2 + \frac{1}{p} |\nabla z|^p + W^{\text{el}}(c, \epsilon(u), z) + W^{\text{ch}}(c) \right) \mathrm{d}x,$$

$$\widetilde{\mathcal{R}}(z_t) := \int_\Omega \left(-\alpha z_t + \frac{\beta}{2} |z_t|^2 \right) \mathrm{d}x.$$

4.2.1 Regularization

The regularization, we want to consider here, is achieved by adding the term $\varepsilon\Delta\partial_t c$ to the Cahn-Hilliard equation (referred to as viscous Cahn-Hilliard equation [BP05]) and the 4-Laplacian $\varepsilon\mathrm{div}(|\nabla u|^2 \nabla u)$ to the quasi-static equilibrium equation in the model problem (3.27). The regularized energies are given by

$$\mathcal{E}_\varepsilon(c, u, z) := \mathcal{E}(c, u, z) + \varepsilon \int_\Omega \frac{1}{4}|\nabla u|^4 \,\mathrm{d}x,$$

$$\widetilde{\mathcal{E}}_\varepsilon(c, u, z) := \widetilde{\mathcal{E}}(c, u, z) + \varepsilon \int_\Omega \frac{1}{4}|\nabla u|^4 \,\mathrm{d}x.$$

The classical formulation of the regularized problem for $\varepsilon > 0$ now reads as

$$\partial_t c = \Delta\mu, \tag{4.6a}$$

$$\mu = -\Delta c + \partial_c W^{\mathrm{ch}}(c) + \partial_c W^{\mathrm{el}}(c, \epsilon(u), z) + \varepsilon\partial_t c, \tag{4.6b}$$

$$\mathrm{div}(\partial_e W^{\mathrm{el}}(c, \epsilon(u), z)) + \varepsilon\mathrm{div}(|\nabla u|^2 \nabla u) = 0, \tag{4.6c}$$

$$0 = -\Delta_p z + \partial_z W^{\mathrm{el}}(c, \epsilon(u), z) + \zeta - \alpha + \beta\partial_t z + \varrho, \tag{4.6d}$$

with $\zeta \in \partial I_{[0,\infty)}(z)$ and $\varrho \in I_{(-\infty,0]}(\partial_t z)$. In the following, we motivate a formulation for weak solutions of this regularized system admissible for functions c, u, z and μ with

$$\begin{cases} c \in L^\infty(0, T; H^1(\Omega)) \cap H^1(0, T; L^2(\Omega)), & u \in L^\infty(0, T; W^{1,4}(\Omega; \mathbb{R}^n)), \\ z \in L^\infty(0, T; W^{1,p}(\Omega)) \cap H^1(0, T; L^2(\Omega)), & \mu \in L^2(0, T; H^1(\Omega)) \end{cases} \tag{4.7}$$

and $c(0) = c^0$, $u|_{(\Gamma_{\mathrm{D}})_T} = b|_{(\Gamma_{\mathrm{D}})_T}$, $z(0) = z^0$, $z \geq 0$ and $\partial_t z \leq 0$ a.e. in Ω_T. We will use the notation $q := (c, u, z)$.

For every $t \in [0, T]$, equation (4.6a) can be translated with the boundary conditions (3.28) in a weak formulation as follows:

$$\int_\Omega (\partial_t c(t))\zeta \,\mathrm{d}x = -\int_\Omega \nabla\mu(t) \cdot \nabla\zeta \,\mathrm{d}x \tag{4.8}$$

for all $\zeta \in H^1(\Omega)$ and

$$\int_\Omega \mu(t)\zeta \,\mathrm{d}x = \int_\Omega \left(\nabla c(t) \cdot \nabla\zeta + \partial_c W^{\mathrm{ch}}(c(t))\zeta + \partial_c W^{\mathrm{el}}(c(t), \epsilon(u(t)), z(t))\zeta + \varepsilon(\partial_t c(t))\zeta\right) \,\mathrm{d}x \tag{4.9}$$

for all $\zeta \in H^1(\Omega)$. In the same spirit, we rewrite (4.6b) as

$$\int_\Omega \left(\partial_e W^{\mathrm{el}}(c(t), \epsilon(u(t)), z(t)) : \epsilon(\zeta) + \varepsilon|\nabla u(t)|^2 \nabla u(t) : \nabla\zeta\right) \,\mathrm{d}x = 0 \tag{4.10}$$

for all $\zeta \in W^{1,4}_{\Gamma_{\mathrm{D}}}(\Omega; \mathbb{R}^n)$ by using the symmetry condition

$$\partial_e W^{\mathrm{el}}(c, e, z) = (\partial_e W^{\mathrm{el}}(c, e, z))^t \quad \text{for } e \in \mathbb{R}^{n\times n}_{\mathrm{sym}}, \ c, z \in \mathbb{R},$$

following from the assumptions (4.2). Note that $|\nabla u|^2 \nabla u \in L^\infty(0,T; L^{4/3}(\Omega; \mathbb{R}^{n\times n}))$ due to (4.7). The differential inclusion (4.6c) is equivalent to

$$0 = \mathrm{d}_z \widetilde{\mathcal{E}}_\varepsilon(q(t)) + r(t) + \mathrm{d}_{\dot{z}} \widetilde{\mathcal{R}}(\partial_t z(t)) + s(t)$$

with some $r(t) \in \partial I_{W_+^{1,p}(\Omega)}(z(t))$ and $s(t) \in \partial I_{W_-^{1,p}(\Omega)}(\partial_t z(t))$. This can be expressed to the following system of variational inequalities:

$$I_{W_-^{1,p}(\Omega)}(\partial_t z(t)) - \Big\langle \mathrm{d}_z \widetilde{\mathcal{E}}_\varepsilon(q(t)) + r(t) + \mathrm{d}_{\dot{z}}\widetilde{\mathcal{R}}(\partial_t z(t)), \zeta - \partial_t z(t) \Big\rangle \le I_{W_-^{1,p}(\Omega)}(\zeta)$$

$$I_{W_+^{1,p}(\Omega)}(z(t)) + \langle r(t), \zeta - z(t)\rangle \le I_{W_+^{1,p}(\Omega)}(\zeta)$$

$$\text{for all } \zeta \in W^{1,p}(\Omega).$$

Here, $\langle \cdot, \cdot \rangle$ denotes the dual pairing between $(W^{1,p}(\Omega))^*$ and $W^{1,p}(\Omega)$. Recall that $W_\pm^{1,p}(\Omega)$ is given by $\{f \in W^{1,p}(\Omega) \,|\, f \gtrless 0\}$. This system is, in turn, equivalent to the inequality system

$$z(t) \ge 0 \text{ and } \partial_t z(t) \le 0, \tag{4.11a}$$

$$- \Big\langle \mathrm{d}_z \widetilde{\mathcal{E}}_\varepsilon(q(t)) + r(t) + \mathrm{d}_{\dot{z}}\widetilde{\mathcal{R}}(\partial_t z(t)), \partial_t z(t) \Big\rangle \ge 0, \tag{4.11b}$$

$$\Big\langle \mathrm{d}_z \widetilde{\mathcal{E}}_\varepsilon(q(t)) + r(t) + \mathrm{d}_{\dot{z}}\widetilde{\mathcal{R}}(\partial_t z(t)), \zeta \Big\rangle \ge 0 \quad \text{for } \zeta \in W_-^{1,p}(\Omega), \tag{4.11c}$$

$$\langle r(t), \zeta - z(t)\rangle \le 0 \quad \text{for } \zeta \in W_+^{1,p}(\Omega). \tag{4.11d}$$

Due to the lack of regularity of q, (4.11b) cannot be justified rigorously. To overcome this difficulty, we use a formal calculation originating from energetic formulations introduced in [MT99].

Proposition 4.2.1 (Energetic characterization)
Let $(c,u,z,\mu) \in \mathcal{C}^2(\overline{\Omega_T}) \times \mathcal{C}^2(\overline{\Omega_T}; \mathbb{R}^n) \times \mathcal{C}^2(\overline{\Omega_T}) \times \mathcal{C}^2(\overline{\Omega_T})$ be a smooth solution of (4.6a)-(4.6c) with the initial-boundary conditions (3.28). Then, the following two conditions are equivalent (set $q := (c,u,z)$):

(i) (4.11b) with $r(t) \in \partial I_{W_+^{1,p}(\Omega)}(z(t))$ for all $t \in [0,T]$,

(ii) for all $0 \le t_1 \le t_2 \le T$:

$$\mathcal{E}_\varepsilon(q(t_2)) - \mathcal{E}_\varepsilon(q(t_1)) + \int_{t_1}^{t_2} \langle \mathrm{d}_{\dot{z}}\widetilde{\mathcal{R}}(\partial_t z), \partial_t z \rangle \, \mathrm{d}s + \int_{t_1}^{t_2} \int_\Omega \left(|\nabla \mu|^2 + \varepsilon |\partial_t c|^2 \right) \mathrm{d}x \, \mathrm{d}s$$

$$\le \int_{t_1}^{t_2} \int_\Omega \partial_e W^{\mathrm{el}}(c, \epsilon(u), z) : \epsilon(\partial_t b) \, \mathrm{d}x \, \mathrm{d}s + \varepsilon \int_{t_1}^{t_2} \int_\Omega |\nabla u|^2 \nabla u : \nabla \partial_t b \, \mathrm{d}x \, \mathrm{d}s.$$

$$\tag{4.12}$$

Proof. We first show for all $t \in [0,T]$:

$$\langle r, \partial_t z(t)\rangle = 0 \text{ for all } r \in \partial I_{W_+^{1,p}(\Omega)}(z(t)). \tag{4.13}$$

The inequality $0 \leq \langle r, \partial_t z(t) \rangle$ follows directly from (4.11d) by putting $\zeta = z(t) - \partial_t z(t)$. The '$\geq$' - part can be shown by an approximation argument. Applying Lemma 2.3.16 with $f_M = z(t)$ and $f = z(t)$ and $\zeta = -\partial_t z(t)$, we obtain a sequence $\{\zeta_M\} \subseteq W_+^{1,p}(\Omega)$ and constants $\nu_M > 0$ such that $-\zeta_M \to \partial_t z(t)$ in $W^{1,p}(\Omega)$ as $M \to \infty$ and $0 \leq z(t) - \nu_M \zeta_M$ a.e. in Ω for all $M \in \mathbb{N}$. Testing (4.11d) with $\zeta = z(t) - \nu_M \zeta_M$ shows $\langle r, -\zeta_M \rangle \leq 0$. Passing to $M \to \infty$ gives $\langle r, \partial_t z(t) \rangle \leq 0$.

To $(ii) \Rightarrow (i)$: We remark that (4.9) and (4.10) can be written in the following form:

$$\int_\Omega \left(\mu(t)\zeta_1 - \varepsilon(\partial_t c(t))\zeta_1 \right) dx = \langle d_c \widetilde{\mathcal{E}}_\varepsilon(q(t)), \zeta_1 \rangle, \tag{4.14a}$$

$$\langle d_u \widetilde{\mathcal{E}}_\varepsilon(q(t)), \zeta_2 \rangle = 0, \tag{4.14b}$$

for all $t \in [0,T]$, all $\zeta_1 \in H^1(\Omega)$ and all $\zeta_2 \in W_{\Gamma_D}^{1,4}(\Omega; \mathbb{R}^n)$.

Let $t_0 \in [0,T)$. It follows

$$\frac{\mathcal{E}_\varepsilon(q(t_0 + h)) - \mathcal{E}_\varepsilon(q(t_0))}{h} + \int_{t_0}^{t_0+h} \langle d_{\dot{z}} \widetilde{\mathcal{R}}(\partial_t z), \partial_t z \rangle \, dt + \int_{t_0}^{t_0+h} \int_\Omega \left(|\nabla \mu|^2 + \varepsilon|\partial_t c|^2 \right) dx \, dt$$

$$\leq \fint_{t_0}^{t_0+h} \int_\Omega \partial_e W^{\mathrm{el}}(c, \epsilon(u), z) : \epsilon(\partial_t b) \, dx \, dt + \varepsilon \fint_{t_0}^{t_0+h} \int_\Omega |\nabla u|^2 \nabla u : \nabla \partial_t b \, dx \, dt.$$

Letting $h \to 0^+$ gives

$$\frac{d}{dt} \widetilde{\mathcal{E}}_\varepsilon(q(t_0)) + \langle d_{\dot{z}} \widetilde{\mathcal{R}}(\partial_t z(t_0)), \partial_t z(t_0) \rangle + \int_\Omega \left(|\nabla \mu(t_0)|^2 + \varepsilon|\partial_t c(t_0)|^2 \right) dx$$

$$\leq \int_\Omega \partial_e W^{\mathrm{el}}(c(t_0), \epsilon(u(t_0)), z(t_0)) : \epsilon(\partial_t b(t_0)) dx$$

$$+ \varepsilon \int_\Omega |\nabla u(t_0)|^2 \nabla u(t_0) : \nabla \partial_t b(t_0) \, dx$$

$$= \langle d_u \widetilde{\mathcal{E}}_\varepsilon(q(t_0)), \partial_t b(t_0) \rangle.$$

Using the chain rule and (4.8)-(4.10) yield

$$\frac{d}{dt} \widetilde{\mathcal{E}}_\varepsilon(q(t_0)) = \underbrace{\langle d_u \widetilde{\mathcal{E}}_\varepsilon(q(t_0)), \partial_t u(t_0) \rangle}_{\text{apply (4.14b)}} + \underbrace{\langle d_c \widetilde{\mathcal{E}}_\varepsilon(q(t_0)), \partial_t c(t_0) \rangle}_{\text{apply (4.14a) and (4.8)}} + \langle d_z \widetilde{\mathcal{E}}_\varepsilon(q(t_0)), \partial_t z(t_0) \rangle$$

$$= \langle d_u \widetilde{\mathcal{E}}_\varepsilon(q(t_0)), \partial_t b(t_0) \rangle + \int_\Omega \left(-|\nabla \mu(t_0)|^2 - \varepsilon|\partial_t c(t_0)|^2 \right) dx$$

$$+ \langle d_z \widetilde{\mathcal{E}}_\varepsilon(q(t_0)), \partial_t z(t_0) \rangle.$$

In consequence, property (i) follows together with (4.13). The case $t_0 = T$ can be derived similarly by considering the difference quotient of t_0 and $t_0 - h$.

To $(i) \Rightarrow (ii)$:   This implication follows from the relation

$$\mathcal{E}_\varepsilon(q(t_2)) - \mathcal{E}_\varepsilon(q(t_1)) = \int_{t_1}^{t_2} \frac{\mathrm{d}}{\mathrm{d}t} \widetilde{\mathcal{E}}_\varepsilon(q(t)) \, \mathrm{d}t$$

as well as the equations (4.8)-(4.10) and (4.13).   □

Remark 4.2.2   *(i) In the rate-independent case $\beta = 0$ and for convex $\mathcal{E}_\varepsilon$ with respect to z, condition (4.11c) can be characterized by a stability condition which reads as*

$$\mathcal{E}_\varepsilon(c(t), u(t), z(t)) \leq \mathcal{E}_\varepsilon(c(t), u(t), \zeta) + \mathcal{R}(\zeta - z(t)) \tag{4.15}$$

for all $t \in [0, T]$ and all test functions $\zeta \in W_+^{1,p}(\Omega)$. Thereby, (4.12) and (4.15) give an equivalent description of the differential inclusion (4.6d) for smooth solutions. This concept of solutions is referred to as global energetic solutions and was introduced in [MT99]. We emphasize that the damage variable z in the rate-independent case $\beta = 0$ is a function of bounded variation and is allowed to exhibit jumps. For a comprehensive introduction, we refer to [AFP00]. To tackle rate-dependent systems and non-convexity of $\mathcal{E}_\varepsilon$ with respect to z, we cannot use formulation (4.15) (cf. [MRS12, MRZ10]).

(ii) For smooth solutions (c, u, z, μ), satisfying (4.8)-(4.10), the energy inequality (4.12) and the variational inequality (4.11c), we even obtain the following energy balance:

$$\mathcal{E}_\varepsilon(q(t_2)) + \int_{t_1}^{t_2} \langle \mathrm{d}_{\dot{z}} \widetilde{\mathcal{R}}(\partial_t z), \partial_t z \rangle \, \mathrm{d}s + \int_{t_1}^{t_2} \int_\Omega (|\nabla \mu|^2 + \varepsilon |\partial_t c|^2) \, \mathrm{d}x \, \mathrm{d}s$$

$$= \mathcal{E}_\varepsilon(q(t_1)) + \int_{t_1}^{t_2} \int_\Omega \partial_e W^{\mathrm{el}}(c, \epsilon(u), z) : \epsilon(\partial_t b) \, \mathrm{d}x \, \mathrm{d}s$$

$$+ \varepsilon \int_{t_1}^{t_2} \int_\Omega |\nabla u|^2 \nabla u : \nabla \partial_t b \, \mathrm{d}x \, \mathrm{d}s$$

for all $0 \leq t_1 \leq t_2 \leq T$.

This motivates the definition of a solution in the following sense:

Definition 4.2.3 (Weak solution for the regularized system (4.6),(3.28)**)**
A 4-tuple (c, u, z, μ) is called a weak solution of the viscous system (4.6) with initial-boundary data (3.28) if it satisfies the following conditions:

(i) the functions are in the following spaces:

$$c \in L^\infty(0, T; H^1(\Omega)) \cap H^1(0, T; L^2(\Omega)), \ c(0) = c^0,$$

$$u \in L^\infty(0, T; W^{1,4}(\Omega; \mathbb{R}^n)), \ u|_{(\Gamma_{\mathrm{D}})_T} = b|_{(\Gamma_{\mathrm{D}})_T},$$

$$z \in L^\infty(0, T; W^{1,p}(\Omega)) \cap H^1(0, T; L^2(\Omega)), \ z(0) = z^0, \ z \geq 0, \ \partial_t z \leq 0 \text{ a.e. in } \Omega_T,$$

$$\mu \in L^2(0, T; H^1(\Omega)),$$

(ii) for all $\zeta \in H^1(\Omega)$ and for a.e. $t \in (0,T)$:

$$\int_\Omega (\partial_t c(t))\zeta \, dx = -\int_\Omega \nabla \mu(t) \cdot \nabla \zeta \, dx, \tag{4.16}$$

(iii) for all $\zeta \in H^1(\Omega)$ and for a.e. $t \in (0,T)$:

$$\int_\Omega \mu(t)\zeta \, dx = \int_\Omega \nabla c(t) \cdot \nabla \zeta \, dx$$
$$+ \int_\Omega \big(\partial_c W^{\mathrm{ch}}(c(t)) + \partial_c W^{\mathrm{el}}(c(t), \epsilon(u(t)), z(t)) + \varepsilon(\partial_t c(t))\big)\zeta \, dx, \tag{4.17}$$

(iv) for all $\zeta \in W_{\Gamma_D}^{1,4}(\Omega; \mathbb{R}^n)$ and for a.e. $t \in (0,T)$:

$$\int_\Omega \big(\partial_e W^{\mathrm{el}}(c(t), \epsilon(u(t)), z(t)) : \epsilon(\zeta) + \varepsilon|\nabla u(t)|^2 \nabla u(t) : \nabla \zeta\big) \, dx = 0, \tag{4.18}$$

(v) for all $\zeta \in W_-^{1,p}(\Omega)$ and for a.e. $t \in (0,T)$:

$$0 \le \int_\Omega |\nabla z(t)|^{p-2} \nabla z(t) \cdot \nabla \zeta \, dx$$
$$+ \int_\Omega \big(\partial_z W^{\mathrm{el}}(c(t), \epsilon(u(t)), z(t)) - \alpha + \beta(\partial_t z(t)) + r(t)\big)\zeta \, dx, \tag{4.19}$$

where $r \in L^1(\Omega_T) \subset L^1\big(0, T; (W^{1,p}(\Omega))^*\big)$ satisfies $r(t) \in \partial I_{W_+^{1,p}(\Omega)}(z(t))$ for a.e.
$t \in (0,T)$, i.e., for all $\zeta \in W_+^{1,p}(\Omega)$ and for a.e. $t \in (0,T)$:

$$\int_\Omega r(t)(\zeta - z(t)) \le 0, \tag{4.20}$$

(vi) energy inequality for a.e. $0 \le t_1 \le t_2 \le T$:

$$\mathcal{E}_\varepsilon(c(t_2), u(t_2), z(t_2)) + \int_{t_1}^{t_2} \int_\Omega \big(-\alpha\partial_t z + \beta|\partial_t z|^2 + |\nabla \mu|^2 + \varepsilon|\partial_t c|^2\big) \, dx \, ds$$
$$\le \mathcal{E}_\varepsilon(c(t_1), u(t_1), z(t_1)) + \int_{t_1}^{t_2} \int_\Omega \big(\partial_e W^{\mathrm{el}}(c, \epsilon(u), z) : \epsilon(\partial_t b) + |\nabla u|^2 \nabla u : \nabla \partial_t b\big) \, dx \, ds. \tag{4.21}$$

Remark 4.2.4 *The regularity properties in (i), the growth conditions (4.2) and standard embedding theorems ensure the existence of all integral terms above:*

- *Due to the growth condition (4.2d) and the embedding $H^1(\Omega) \hookrightarrow L^4(\Omega)$ for c (with space dimension $n \in \{1,2,3\}$), the function $\partial_c W^{\mathrm{el}}(c(t), \epsilon(u(t)), z(t))$ is in $L^2(\Omega)$.*

- *The growth condition (4.2g) and the embedding $H^1(\Omega) \hookrightarrow L^{2^*}(\Omega)$ show that $\partial_c W^{\mathrm{ch}}(c(t))$ is in $L^2(\Omega)$.*

- *The functions $\partial_z W^{\mathrm{el}}(c(t), \epsilon(u(t)), z(t))$ and $r(t)$ are in $L^1(\Omega)$. To see that the corresponding integral terms in (4.19) and (4.20) exist, we have to take the embedding $W^{1,p}(\Omega) \hookrightarrow \mathcal{C}(\overline{\Omega})$ into account. More precisely, we take advantage of $\zeta \in L^\infty(\Omega)$ and $z(t) \in L^\infty(\Omega)$ for a.e. $t \in (0, T)$.*

Theorem 4.2.5 (Existence theorem - viscous problem) *Let the assumptions (4.1) and (4.2) be satisfied and let $b \in W^{1,1}(0, T; W^{1,\infty}(\Omega; \mathbb{R}^n))$, $c^0 \in H^1(\Omega)$, $z^0 \in W^{1,p}(\Omega)$ with $0 \leq z^0 \leq 1$ and a viscosity factor $\varepsilon \in (0, 1)$ be given.*

Then, there exists a weak solution (c, u, z, μ) in the sense of Definition 4.2.3. In addition:

$$r = -\chi_{\{z=0\}}[\partial_z W^{\mathrm{el}}(c, \epsilon(u), z)]^+, \tag{4.22}$$

where $[\cdot]^+$ is defined by $\max\{0, \cdot\}$.

4.2.2   Limit problem

Our main aim in this work is to establish an existence result for the system given in Definition 3.4.1. In the same fashion as in Section 4.2.1, we introduce a weak notion of (3.27) as follows.

Definition 4.2.6 (Weak solution for the limit system (3.27)-(3.28))
A 4-tuple (u, c, z, μ) is called a weak solution of the limit system (3.27) and with initial-boundary conditions (3.28) if it satisfies the following properties:

(i) the functions are in the following spaces:

$$c \in L^\infty(0, T; H^1(\Omega)) \cap H^1(0, T; (H^1(\Omega))^*),\ c(0) = c^0,$$
$$u \in L^\infty(0, T; H^1(\Omega; \mathbb{R}^n)),\ u|_{(\Gamma_{\mathrm{D}})_T} = b|_{(\Gamma_{\mathrm{D}})_T},$$
$$z \in L^\infty(0, T; W^{1,p}(\Omega)) \cap H^1(0, T; L^2(\Omega)),\ z(0) = z^0,\ z \geq 0,\ \partial_t z \leq 0\ \text{a.e. in } \Omega_T,$$
$$\mu \in L^2(0, T; H^1(\Omega)).$$

(ii) for all $\zeta \in L^2(0, T; H^1(\Omega))$ with $\partial_t \zeta \in L^2(\Omega_T)$ and $\zeta(T) = 0$:

$$\int_{\Omega_T} (c(t) - c^0)\partial_t \zeta \, \mathrm{d}x \, \mathrm{d}t = \int_{\Omega_T} \nabla \mu(t) \cdot \nabla \zeta \, \mathrm{d}x \, \mathrm{d}t,$$

(iii) for all $\zeta \in H^1(\Omega)$ and for a.e. $t \in (0, T)$:

$$\int_\Omega \mu(t)\zeta \, \mathrm{d}x = \int_\Omega \Big(\nabla c(t) \cdot \nabla \zeta + (\partial_c W^{\mathrm{ch}}(c(t)) + \partial_c W^{\mathrm{el}}(c(t), \epsilon(u(t)), z(t)))\zeta \Big) \, \mathrm{d}x,$$

(iv) for all $\zeta \in H^1_{\Gamma_D}(\Omega; \mathbb{R}^n)$ and for a.e. $t \in (0, T)$:

$$\int_\Omega \partial_e W^{el}(c(t), \epsilon(u(t)), z(t)) : \epsilon(\zeta) \, dx = 0,$$

(v) for all $\zeta \in W^{1,p}_-(\Omega)$ and for a.e. $t \in (0, T)$:

$$0 \le \int_\Omega |\nabla z(t)|^{p-2} \nabla z(t) \cdot \nabla \zeta \, dx$$
$$+ \int_\Omega \left(\partial_z W^{el}(c(t), \epsilon(u(t)), z(t)) - \alpha + \beta(\partial_t z(t)) + r(t) \right) \zeta \, dx,$$

where $r \in L^1(\Omega_T) \subset L^1(0, T; (W^{1,p}(\Omega))^*)$ satisfies for all $\zeta \in W^{1,p}_+(\Omega)$ and for a.e. $t \in (0, T)$:

$$\int_\Omega r(t)(\zeta - z(t)) \le 0,$$

(vi) for a.e. $0 \le t_1 \le t_2 \le T$:

$$\mathcal{E}(c(t_2), u(t_2), z(t_2)) + \int_{t_1}^{t_2} \int_\Omega \left(-\alpha \partial_t z + \beta |\partial_t z|^2 + |\nabla \mu|^2 \right) dx \, ds$$
$$\le \mathcal{E}(c(t_1), u(t_1), z(t_1)) + \int_{t_1}^{t_2} \int_\Omega \partial_e W^{el}(c, \epsilon(u), z) : \epsilon(\partial_t b) \, dx \, ds.$$

Theorem 4.2.7 (Existence theorem - limit problem) *Let the assumptions (4.1) and (4.2) be satisfied and let $b \in W^{1,1}(0, T; W^{1,\infty}(\Omega; \mathbb{R}^n))$, $c^0 \in H^1(\Omega)$ and $z^0 \in W^{1,p}(\Omega)$ with $0 \le z^0 \le 1$ be given.*

Then, there exists a weak solution (c, u, z, μ) in the sense of Definition 4.2.6. In addition:

$$r = -\chi[\partial_z W^{el}(c, \epsilon(u), z)]^+,$$

where the function $\chi \in L^\infty(\Omega_T)$ satisfies $\chi = 0$ in $\{z > 0\}$ and $0 \le \chi \le 1$ in $\{z = 0\}$.

4.3 Proofs of the existence theorems

4.3.1 Existence of weak solutions for the regularized system

This section is aimed to prove Theorem 4.2.5. The initial displacement u^0_ε is chosen to be a minimizer of the functional $u \mapsto \mathcal{E}_\varepsilon(c^0, \epsilon(u), z^0)$ defined on the space $W^{1,4}(\Omega)$ with the constraint $u|_{\Gamma_D} = b(0)|_{\Gamma_D}$.

To proceed, we introduce the following spaces:

$$V_0 := \left\{ \zeta \in H^1(\Omega) \,\Big|\, \int_\Omega \zeta \, dx = 0 \right\}, \tag{4.23a}$$

$$\tilde{V}_0 := \left\{ \zeta \in (H^1(\Omega))^* \mid \langle \zeta, 1 \rangle_{(H^1)^* \times H^1} = 0 \right\}. \tag{4.23b}$$

This permits us to define the operator $(-\Delta)^{-1} : \tilde{V}_0 \to V_0$ as the inverse of the operator $-\Delta : V_0 \to \tilde{V}_0$, $u \mapsto \langle \nabla u, \nabla \cdot \rangle_{L^2(\Omega)}$. The space $\tilde{V}_0$ will be endowed with the scalar product $\langle u, v \rangle_{\tilde{V}_0} := \langle \nabla(-\Delta)^{-1}u, \nabla(-\Delta)^{-1}v \rangle_{L^2(\Omega)}$.

We now apply an implicit Euler scheme for the system (4.6). The discretization fineness is given by $\tau := \frac{T}{M}$, where $M \in \mathbb{N}$. We set $q^0_{M,\varepsilon} := (c^0, u^0_\varepsilon, z^0)$ and construct $q^m_{M,\varepsilon} := (c^m_{M,\varepsilon}, u^m_{M,\varepsilon}, z^m_{M,\varepsilon})$ for $m \in \{1, \ldots, M\}$ recursively by considering the functional

$$\mathbb{E}^m_{M,\varepsilon}(c, u, z) := \tilde{\mathcal{E}}_\varepsilon(c, u, z) + \tilde{\mathcal{R}}\left(\frac{z - z^{m-1}_{M,\varepsilon}}{\tau}\right)\tau + \frac{1}{2\tau}\|c - c^{m-1}_{M,\varepsilon}\|^2_{\tilde{V}_0} + \frac{\varepsilon}{2\tau}\|c - c^{m-1}_{M,\varepsilon}\|^2_{L^2(\Omega)}.$$

The set of admissible states for $\mathbb{E}^m_{M,\varepsilon}$ is

$$\mathfrak{Q}^m_{M,\varepsilon} := \Big\{ (c, u, z) \in H^1(\Omega) \times W^{1,4}(\Omega; \mathbb{R}^n) \times W^{1,p}(\Omega)$$

$$\text{with } u|_{\Gamma_D} = b(m\tau)|_{\Gamma_D}, \int_\Omega (c - c^0)\, dx = 0 \text{ and } 0 \leq z \leq z^{m-1}_{M,\varepsilon} \text{ in } \Omega \Big\}.$$

A minimization problem (without the damage variable) for the functional

$$\mathbb{E}^m_{M,\varepsilon}(c, u, z) = \mathbb{E}^m_{M,\varepsilon}(c, u) = \int_\Omega \left(\frac{1}{2}|\nabla c|^2 + W^{\mathrm{ch}}(c) + W^{\mathrm{el}}(c, \epsilon(u))\right) dx + \frac{1}{2\tau}\|c - c^{m-1}_{M,\varepsilon}\|^2_{\mathrm{L}}$$

containing a weighted $(H^1(\Omega, \mathbb{R}^n))^*$-scalar product $\langle \cdot, \cdot \rangle_{\mathrm{L}}$ has been considered in [Gar00]. However, due to the additional internal variable z, the passage to $M \to \infty$ becomes much more involved.

In the following, we will omit the ε-dependence in the notation since $\varepsilon \in (0,1)$ is fixed until Section 4.3.2.

Lemma 4.3.1 *The functional $\mathbb{E}^m_M$ has a minimizer $q^m_M \in \mathfrak{Q}^m_M$.*

Proof. The existence is shown by direct methods in the calculus of variations. We can immediately see that $\mathfrak{Q}^m_M$ is closed with respect to the weak topology in $H^1(\Omega) \times W^{1,4}(\Omega; \mathbb{R}^n) \times W^{1,p}(\Omega)$. Furthermore, we need to show coercivity and sequentially weakly lower semi-continuity of $\mathbb{E}^m_M$ defined on $\mathfrak{Q}^m_M$.

(i) *Coercivity.* We have the estimate

$$\mathbb{E}^m_M(c, u, z) \geq \frac{1}{2}\|\nabla c\|^2_{L^2(\Omega)} + \frac{1}{p}\|\nabla z\|^p_{L^p(\Omega)} + \frac{\varepsilon}{4}\|\nabla u\|^4_{L^4(\Omega)}.$$

Therefore, given a sequence $\{c_k, u_k, z_k\}_{k \in \mathbb{N}}$ in $\mathfrak{Q}^m_M$ with the boundedness property $\mathbb{E}^m_M(c_k, u_k, z_k) < C$ for all $k \in \mathbb{N}$, we obtain the boundedness of u_k in $W^{1,4}(\Omega)$ by Poincaré's inequality (u_k has fixed boundary data on Γ_D), the boundedness of c_k in $H^1(\Omega)$ by Poincaré's inequality ($\int_\Omega c_k\, dx$ is conserved) and the boundedness of z_k in $W^{1,p}(\Omega)$ by also considering the restriction $0 \leq z_k \leq 1$ in Ω.

(ii) *Sequentially weakly lower semi-continuity.* All terms in $\mathbb{E}_M^m$ except $\int_\Omega W^{\mathrm{ch}}(c)\,\mathrm{d}x$ and $\int_\Omega W^{\mathrm{el}}(c, \epsilon(u), z)\,\mathrm{d}x$ are convex and continuous and therefore sequentially weakly l.s.c.. Now let $(c_k, u_k, z_k) \rightharpoonup (c, u, z)$ be a weakly converging sequence in $\mathcal{Q}_M^m$. In particular, $z_k \to z$ in $\mathcal{C}(\overline{\Omega})$ and $c_k \to c$ in $L^r(\Omega)$ as $k \to \infty$ for all $1 \le r < 2^\star$ and $c_k \to c$ a.e. in Ω for a subsequence. Lebesgue's generalized convergence theorem yields $\int_\Omega W^{\mathrm{ch}}(c_k)\,\mathrm{d}x \to \int_\Omega W^{\mathrm{ch}}(c)\,\mathrm{d}x$ using (4.2g). The remaining term can be treated by employing the uniform convexity of $W^{\mathrm{el}}(c, \cdot, z)$ (see (4.2a)):

$$\int_\Omega \left(W^{\mathrm{el}}(c_k, \epsilon(u_k), z_k) - W^{\mathrm{el}}(c, \epsilon(u), z) \right)\mathrm{d}x$$
$$= \int_\Omega \left(W^{\mathrm{el}}(c_k, \epsilon(u), z_k) - W^{\mathrm{el}}(c, \epsilon(u), z) \right)\mathrm{d}x$$
$$+ \int_\Omega \left(W^{\mathrm{el}}(c_k, \epsilon(u_k), z_k) - W^{\mathrm{el}}(c_k, \epsilon(u), z_k) \right)\mathrm{d}x$$
$$\ge \underbrace{\int_\Omega \left(W^{\mathrm{el}}(c_k, \epsilon(u), z_k) - W^{\mathrm{el}}(c, \epsilon(u), z) \right)\mathrm{d}x}_{\to 0 \text{ by Lebesgue's gen. conv. theorem and (4.2c)}}$$
$$+ \int_\Omega \partial_e W^{\mathrm{el}}(c_k, \epsilon(u), z_k) : (\epsilon(u_k) - \epsilon(u))\,\mathrm{d}x.$$

The second term converges to 0 because of $\partial_e W^{\mathrm{el}}(c_k, \epsilon(u), z_k) \to \partial_e W^{\mathrm{el}}(c, \epsilon(u), z)$ in $L^2(\Omega)$ (by Lebesgue's generalized convergence theorem and (4.3)) and $\epsilon(u_k) - \epsilon(u) \rightharpoonup 0$ in $L^2(\Omega)$.

Thus there exists $(c_M^m, u_M^m, z_M^m) \in \mathcal{Q}_M^m$ such that we obtain the desired property $\mathbb{E}_M^m(c_M^m, u_M^m, z_M^m) = \inf_{(c,u,z)\in\mathcal{Q}_M^m} \mathbb{E}_M^m(c, u, z)$. □

The minimizers q_M^m for $m \in \{0, \dots, M\}$ are used to construct approximate solutions $q_M := (c_M, u_M, z_M)$, $q_M^- := (c_M^-, u_M^-, z_M^-)$ and $\widehat{q}_M := (\widehat{c}_M, \widehat{u}_M, \widehat{z}_M)$ to our viscous problem by a piecewise constant and linear interpolation in time, respectively. More precisely,

$$q_M(t) := q_M^m,$$
$$\widehat{q}_M(t) := \beta q_M^m + (1 - \beta) q_M^{m-1}$$

with $t \in ((m-1)\tau, m\tau]$ and $\beta = \frac{t - (m-1)\tau}{\tau}$. The retarded function q_M^- is set to

$$q_M^-(t) := \begin{cases} q_M(t - \tau), & \text{if } t \in [\tau, T], \\ (c^0, u_\epsilon^0, z^0), & \text{if } t \in [0, \tau). \end{cases}$$

The functions b_M and b_M^- are analogously defined adopting the notation $b_M^m := b(m\tau)$. Furthermore, the discrete chemical potential is given by (note that $\partial_t \widehat{c}_M(t) \in V_0$)

$$\mu_M(t) := -(-\Delta)^{-1} \left(\partial_t \widehat{c}_M(t) \right) + \lambda_M(t) \tag{4.24}$$

with the Lagrange multiplier λ_M (originating from mass conservation):

$$\lambda_M(t) := \int_\Omega \left(\partial_c W^{\text{ch}}(c_M(t)) + \partial_c W^{\text{el}}(c_M(t), \epsilon(u_M(t)), z_M(t)) \right) \mathrm{d}x. \tag{4.25}$$

The discretization of the time variable t will be expressed by the functions

$$d_M(t) := \min\{m\tau \,|\, m \in \mathbb{N}_0 \text{ and } m\tau \geq t\},$$
$$d_M^-(t) := \min\{(m-1)\tau \,|\, m \in \mathbb{N}_0 \text{ and } m\tau \geq t\}.$$

The following lemma clarifies why the functions q_M, q_M^- and $\widehat{q}_M$ are approximate solutions to our problem.

Lemma 4.3.2 (Euler-Lagrange equations and energy inequality)
The following properties are satisfied:

(i) for all $\zeta \in H^1(\Omega)$ and for all $t \in (0, T)$:

$$\int_\Omega (\partial_t \widehat{c}_M(t)) \zeta \, \mathrm{d}x = -\int_\Omega \nabla \mu_M(t) \cdot \nabla \zeta \, \mathrm{d}x, \tag{4.26}$$

(ii) for all $\zeta \in H^1(\Omega)$ and for all $t \in (0, T)$:

$$\int_\Omega \mu_M(t) \zeta \, \mathrm{d}x = \int_\Omega \left(\nabla c_M(t) \cdot \nabla \zeta + \partial_c W^{\text{ch}}(c_M(t)) \zeta \right) \mathrm{d}x$$
$$+ \int_\Omega \left(\partial_c W^{\text{el}}(c_M(t), \epsilon(u_M(t)), z_M(t)) + \varepsilon(\partial_t \widehat{c}_M(t)) \right) \zeta \, \mathrm{d}x, \tag{4.27}$$

(iii) for all $\zeta \in W_{\Gamma_D}^{1,4}(\Omega; \mathbb{R}^n)$ and for all $t \in (0, T)$:

$$0 = \int_\Omega \left(\partial_e W^{\text{el}}(c_M(t), \epsilon(u_M(t)), z_M(t)) : \epsilon(\zeta) + \varepsilon |\nabla u_M(t)|^2 \nabla u_M(t) : \nabla \zeta \right) \mathrm{d}x,$$
$$\tag{4.28}$$

(iv) for all $t \in (0, T)$ and for all $\zeta \in W^{1,p}(\Omega)$ such that there exists a constant $\nu > 0$ which satisfies the estimate $0 \leq \nu \zeta + z_M(t) \leq z_M^-(t)$ a.e. in Ω:

$$0 \leq \int_\Omega |\nabla z_M(t)|^{p-2} \nabla z_M(t) \cdot \nabla \zeta \, \mathrm{d}x$$
$$+ \int_\Omega \left(\partial_z W^{\text{el}}(c_M(t), \epsilon(u_M(t)), z_M(t)) - \alpha + \beta(\partial_t \widehat{z}_M(t)) \right) \zeta \, \mathrm{d}x, \tag{4.29}$$

(v) energy inequality for all $t \in (0, T)$:

$$\mathcal{E}_\varepsilon(q_M(t)) + \int_0^{d_M(t)} \mathcal{R}(\partial_t \widehat{z}_M) \, \mathrm{d}s + \int_0^{d_M(t)} \int_\Omega \left(\frac{\varepsilon}{2} |\partial_t \widehat{c}_M|^2 + \frac{1}{2} |\nabla \mu_M|^2 \right) \mathrm{d}x \, \mathrm{d}s$$

$$\leq \mathcal{E}_\varepsilon(q^0) + \int_0^{d_M(t)} \int_\Omega \partial_e W^{\mathrm{el}}(c_M^-, \epsilon(u_M^- + b - b_M^-), z_M^-) : \epsilon(\partial_t b) \,\mathrm{d}x \,\mathrm{d}s$$

$$+ \varepsilon \int_0^{d_M(t)} \int_\Omega |\nabla u_M^- + \nabla b - \nabla b_M^-|^2 \nabla(u_M^- + b - b_M^-) : \nabla \partial_t b \,\mathrm{d}x \,\mathrm{d}s.$$

$$(4.30)$$

Proof. By using Lebesgue's generalized convergence theorem, the mean value theorem of differentiability and growth conditions (4.2d)-(4.2g) and (4.3), we obtain the Gâteaux derivatives

$$\mathrm{d}_c \widetilde{\mathcal{E}}_\varepsilon : H^1(\Omega) \times W^{1,4}(\Omega; \mathbb{R}^n) \times W^{1,p}(\Omega) \to (H^1(\Omega))^*,$$

$$\mathrm{d}_u \widetilde{\mathcal{E}}_\varepsilon : H^1(\Omega) \times W^{1,4}(\Omega; \mathbb{R}^n) \times W^{1,p}(\Omega) \to (W^{1,4}(\Omega; \mathbb{R}^n))^*,$$

$$\mathrm{d}_z \widetilde{\mathcal{E}}_\varepsilon : H^1(\Omega) \times W^{1,4}(\Omega; \mathbb{R}^n) \times W^{1,p}(\Omega) \to (W^{1,p}(\Omega))^*$$

of the functional $\widetilde{\mathcal{E}} : H^1(\Omega) \times W^{1,4}(\Omega; \mathbb{R}^n) \times W^{1,p}(\Omega) \to \mathbb{R}$ with respect to c, u and z as (set $q := (c, u, z)$)

for $\zeta \in H^1(\Omega)$:

$$\langle \mathrm{d}_c \widetilde{\mathcal{E}}_\varepsilon(q), \zeta \rangle = \int_\Omega \left(\nabla c \cdot \nabla \zeta + \partial_c W^{\mathrm{ch}}(c)\zeta + \partial_c W^{\mathrm{el}}(c, \epsilon(u), z)\zeta \right) \mathrm{d}x, \qquad (4.31a)$$

for $\zeta \in W^{1,4}(\Omega; \mathbb{R}^n)$:

$$\langle \mathrm{d}_u \widetilde{\mathcal{E}}_\varepsilon(q), \zeta \rangle = \int_\Omega \left(\partial_e W^{\mathrm{el}}(c, \epsilon(u), z) : \epsilon(\zeta) + \varepsilon|\nabla u|^2 \nabla u : \nabla \zeta \right) \mathrm{d}x, \qquad (4.31b)$$

for $\zeta \in W^{1,p}(\Omega)$:

$$\langle \mathrm{d}_z \widetilde{\mathcal{E}}_\varepsilon(q), \zeta \rangle = \int_\Omega \left(|\nabla z|^{p-2} \nabla z \cdot \nabla \zeta + \partial_z W^{\mathrm{el}}(c, \epsilon(u), z)\zeta \right) \mathrm{d}x. \qquad (4.31c)$$

We will exemplarily show the Gâteaux derivation of the elastic energy term in $\widetilde{\mathcal{E}}_\varepsilon$ with respect to u. The other terms can be treated analogously. Now, let $c \in H^1(\Omega)$ and $u, \zeta \in W^{1,4}(\Omega; \mathbb{R}^n)$ and $z \in W^{1,p}(\Omega)$. By the differentiability of W^{el}, we get a.e. in Ω the pointwise convergence

$$\frac{1}{t}\left(W^{\mathrm{el}}(c, \epsilon(u) + t\epsilon(\zeta), z) - W^{\mathrm{el}}(c, \epsilon(u), z) \right) \to \partial_e W^{\mathrm{el}}(c, \epsilon(u), z) : \epsilon(\zeta) \text{ as } t \to 0.$$

Moreover, the following upper estimate holds by applying the mean value theorem of differentiability and (4.3) (with $\xi \in [0, t]$ suitably chosen and $|t| < 1$):

$$\left| \frac{1}{t}\left(W^{\mathrm{el}}(c, \epsilon(u) + t\epsilon(\zeta), z) - W^{\mathrm{el}}(c, \epsilon(u), z) \right) \right| \leq \left| \frac{1}{t} \partial_e W^{\mathrm{el}}(c, \epsilon(u) + \xi\epsilon(\zeta), z) : t\epsilon(\zeta) \right|$$

$$\leq C(|c|^2 + |\epsilon(u)| + |\epsilon(\zeta)| + 1)|\epsilon(\zeta)|.$$

Since the right hand side is an $L^1(\Omega)$ function, Lebesgue's convergence theorem implies as $t \to 0$:

$$\frac{1}{t} \int_\Omega \left(W^{\mathrm{el}}(c, \epsilon(u) + t\epsilon(\zeta), z) - W^{\mathrm{el}}(c, \epsilon(u), z) \right) \mathrm{d}x \to \int_\Omega \partial_e W^{\mathrm{el}}(c, \epsilon(u), z) : \epsilon(\zeta) \,\mathrm{d}x.$$

To (i)-(v):

(i) This follows from (4.24).

(ii) q_M^m fulfills $\langle d_c \mathbb{E}_M^m(q_M^m), \zeta_1 \rangle = 0$ for all $\zeta_1 \in V_0$ and all $m \in \{1, \ldots, M\}$. Therefore,

$$0 = \langle d_c \widetilde{\mathcal{E}}_\varepsilon(q_M(t)), \zeta_1 \rangle + \langle \partial_t \widehat{c}_M(t), \zeta_1 \rangle_{\widetilde{V}_0} + \varepsilon \langle \partial_t \widehat{c}_M(t), \zeta_1 \rangle_{L^2(\Omega)}.$$

On the one hand, definition (4.24) implies

$$\begin{aligned}
\langle \partial_t \widehat{c}_M(t), \zeta_1 \rangle_{\widetilde{V}_0} &= \langle (-\Delta)^{-1} (\partial_t \widehat{c}_M(t)), \zeta_1 \rangle_{L^2(\Omega)} \\
&= \langle -\mu_M(t) + \lambda_M(t), \zeta_1 \rangle_{L^2(\Omega)} \\
&= -\langle \mu_M(t), \zeta_1 \rangle_{L^2(\Omega)}
\end{aligned}$$

and consequently

$$0 = \langle d_c \widetilde{\mathcal{E}}_\varepsilon(q_M(t)), \zeta_1 \rangle - \langle \mu_M(t), \zeta_1 \rangle_{L^2(\Omega)} + \varepsilon \langle \partial_t \widehat{c}_M(t), \zeta_1 \rangle_{L^2(\Omega)} \qquad (4.32)$$

for all $\zeta_1 \in V_0$. On the other hand, definitions (4.24) and (4.25) yield for $\zeta_2 \equiv \widetilde{C}$ with constant $\widetilde{C} \in \mathbb{R}$:

$$\begin{aligned}
&\langle d_c \widetilde{\mathcal{E}}_\varepsilon(q_M(t)), \zeta_2 \rangle - \langle \mu_M(t), \zeta_2 \rangle_{L^2(\Omega)} + \varepsilon \langle \partial_t \widehat{c}_M(t), \zeta_2 \rangle_{L^2(\Omega)} \\
&= \widetilde{C} \mathcal{L}^n(\Omega) \lambda_M(t) + \underbrace{\langle (-\Delta)^{-1} (\partial_t \widehat{c}_M(t)), \zeta_2 \rangle_{L^2(\Omega)}}_{=0} - \underbrace{\langle \lambda_M(t), \zeta_2 \rangle_{L^2(\Omega)}}_{\widetilde{C} \mathcal{L}^n(\Omega) \lambda_M(t)} + 0 \\
&= 0.
\end{aligned} \qquad (4.33)$$

Setting $\zeta_1 = \zeta - f\zeta$ and $\zeta_2 = f\zeta$, inserting (4.31a) into (4.32) and (4.33), and adding (4.32) to (4.33) shows finally (ii) (cf. [Gar00, Lemma 3.2]).

(iii) This property follows from (4.31b) and $0 = \langle d_u \mathbb{E}_M^m(q_M^m), \zeta \rangle = \langle d_u \widetilde{\mathcal{E}}_\varepsilon(q_M^m), \zeta \rangle$ for all $\zeta \in W_{\Gamma_D}^{1,4}(\Omega; \mathbb{R}^n)$.

(iv) By construction, z_M^m minimizes $\mathbb{E}_M^m(u_M^m, c_M^m, \cdot)$ in the space $W^{1,p}(\Omega)$ with the constraints $0 \le z$ and $z - z_M^{m-1} \le 0$ a.e. in Ω. The minimizer is characterized by the following property

$$-\langle d_z \widetilde{\mathcal{E}}_\varepsilon(q_M^m) + d_{\dot{z}} \widetilde{\mathcal{R}}((z_M^m - z_M^{m-1})/\tau), \zeta - z_M^m \rangle_{W^{1,p}} \le 0 \qquad (4.34)$$

for all $\zeta \in W^{1,p}(\Omega)$ with $0 \le \zeta \le z_M^{m-1}$ a.e. in Ω. Now, let the functions $\zeta \in W^{1,p}(\Omega)$ and $\nu > 0$ with $0 \le \nu\zeta + z_M(t) \le z_M^-(t)$ a.e. in Ω be given. Since $\nu > 0$, we obtain from (4.34):

$$-\langle d_z \widetilde{\mathcal{E}}_\varepsilon(q_M(t)), \zeta(t) \rangle_{W^{1,p}} - \langle d_{\dot{z}} \widetilde{\mathcal{R}}(\partial_t \widehat{z}_M(t)), \zeta(t) \rangle_{L^2} \le 0.$$

This and (4.31c) gives (iv).

(v) Testing $\mathbb{E}_M^m$ with $q = (u_M^{m-1} + b_M^m - b_M^{m-1}, c_M^{m-1}, z_M^{m-1})$ and using the chain rule yield:

$$
\mathcal{E}_\varepsilon(q_M^m) + \mathcal{R}\left(\frac{z_M^m - z_M^{m-1}}{\tau}\right)\tau + \frac{1}{2\tau}\|c_M^m - c_M^{m-1}\|_{\tilde{V}_0}^2 + \frac{\varepsilon}{2\tau}\|c_M^m - c_M^{m-1}\|_{L^2(\Omega)}^2
$$

$$
\leq \mathcal{E}_\varepsilon(c_M^{m-1}, u_M^{m-1} + b_M^m - b_M^{m-1}, z_M^{m-1})
$$

$$
= \mathcal{E}_\varepsilon(q_M^{m-1}) + \mathcal{E}_\varepsilon(c_M^{m-1}, u_M^{m-1} + b_M^m - b_M^{m-1}, z_M^{m-1}) - \mathcal{E}_\varepsilon(q_M^{m-1})
$$

$$
= \mathcal{E}_\varepsilon(q_M^{m-1}) + \int_{(m-1)\tau}^{m\tau} \frac{\mathrm{d}}{\mathrm{d}s} \mathcal{E}_\varepsilon(c_M^{m-1}, u_M^{m-1} + b(s) - b_M^{m-1}, z_M^{m-1}) \, \mathrm{d}s
$$

$$
= \mathcal{E}_\varepsilon(q_M^{m-1})
$$

$$
+ \int_{(m-1)\tau}^{m\tau} \int_\Omega \partial_e W^{\mathrm{el}}(c_M^{m-1}, \epsilon(u_M^{m-1} + b(s) - b_M^{m-1}), z_M^{m-1}) : \epsilon(\partial_t b) \, \mathrm{d}x \, \mathrm{d}s
$$

$$
+ \varepsilon \int_{(m-1)\tau}^{m\tau} \int_\Omega |\nabla u_M^{m-1} + \nabla b(s) - \nabla b_M^{m-1}|^2 \nabla(u_M^{m-1} + b(s) - b_M^{m-1}) : \nabla \partial_t b \, \mathrm{d}x \, \mathrm{d}s.
$$

Summing this inequality for $k = 1, \ldots, m$ one gets:

$$
\mathcal{E}_\varepsilon(q_M^m) + \sum_{k=1}^m \tau\left(\mathcal{R}\left(\frac{z_M^k - z_M^{k-1}}{\tau}\right) + \frac{1}{2}\left\|\frac{c_M^k - c_M^{k-1}}{\tau}\right\|_{\tilde{V}_0}^2 + \frac{\varepsilon}{2}\left\|\frac{c_M^k - c_M^{k-1}}{\tau}\right\|_{L^2(\Omega)}^2\right)
$$

$$
\leq \mathcal{E}_\varepsilon(q_\varepsilon^0) + \int_0^{m\tau} \int_\Omega \partial_e W^{\mathrm{el}}(c_M^-, \epsilon(u_M^- + b - b_M^-), z_M^-) : \epsilon(\partial_t b) \, \mathrm{d}x \, \mathrm{d}s
$$

$$
+ \varepsilon \int_0^{m\tau} \int_\Omega |\nabla u_M^- + \nabla b - \nabla b_M^-|^2 \nabla(u_M^- + b - b_M^-) : \nabla \partial_t b \, \mathrm{d}x \, \mathrm{d}s.
$$

Because of $\left\|\frac{c_M^k - c_M^{k-1}}{\tau}\right\|_{\tilde{V}_0}^2 = \|\nabla \mu_M^k\|_{L^2(\Omega)}^2$ by (4.24), the above estimate shows (v). $\qquad \square$

The discrete energy inequality (4.30) gives rise to a-priori estimates for the approximate solutions.

Lemma 4.3.3 (Energy boundedness) *There exists a constant $C > 0$ independent of M, t and ε such that*

$$
\mathcal{E}_\varepsilon(q_M(t)) + \int_0^{d_M(t)} \mathcal{R}(\partial_t \widehat{z}_M) \, \mathrm{d}s + \int_0^{d_M(t)} \int_\Omega \left(\frac{\varepsilon}{2}|\partial_t \widehat{c}_M|^2 + \frac{1}{2}|\nabla \mu_M|^2\right) \mathrm{d}x \, \mathrm{d}s \leq C(\mathcal{E}_\varepsilon(q_\varepsilon^0) + 1).
$$

Proof. Exploiting (4.2e) yields the estimate ($C > 0$ denotes a context-dependent constant independent of M, t and ε):

$$
\int_\Omega \partial_e W^{\mathrm{el}}(c_M^-(s), \epsilon(u_M^-(s) + b(s) - b_M^-(s)), z_M^-(s)) : \epsilon(\partial_t b(s)) \, \mathrm{d}x
$$

$$\leq C\|\nabla\partial_t b(s)\|_{L^\infty(\Omega)}\int_\Omega \left(W^{\mathrm{el}}(c_M^-(s),\epsilon(u_M^-(s)),z_M^-(s))+|\epsilon(b(s)-b_M^-(s))|+1\right)\mathrm{d}x.$$

(4.35)

In addition,

$$\int_\Omega |\nabla u_M^-(s)+\nabla b(s)-\nabla b_M^-(s)|^2\nabla(u_M^-(s)+b(s)-b_M^-(s)):\nabla\partial_t b(s)\,\mathrm{d}x$$

$$\leq C\|\nabla\partial_t b(s)\|_{L^\infty(\Omega)}\int_\Omega |\nabla u_M^-(s)|^3+|\nabla(b(s)-b_M^-(s))|^3\,\mathrm{d}x.$$

(4.36)

To simplify the notation, we define the function:

$$\gamma(t):=$$
$$\begin{cases}\mathcal{E}_\varepsilon(q_M(t))+\int_0^{d_M(t)}\mathcal{R}(\partial_t\widehat{z}_M)\,\mathrm{d}s+\int_0^{d_M(t)}\int_\Omega\left(\frac{\varepsilon}{2}|\partial_t\widehat{c}_M|^2+\frac{1}{2}|\nabla\mu_M|^2\right)\mathrm{d}x\,\mathrm{d}s, & \text{if } t\in[0,T],\\ \mathcal{E}_\varepsilon(q_\varepsilon^0), & \text{if } t\in[-\tau,0).\end{cases}$$

Using (4.35) and (4.36), the discrete energy inequality (4.30) can be estimated as follows:

$$\gamma(t)\leq\mathcal{E}_\varepsilon(q_\varepsilon^0)+C\int_0^{d_M(t)}\|\nabla\partial_t b(s)\|_{L^\infty(\Omega)}\mathcal{E}_\varepsilon(q_M^-(s))\,\mathrm{d}s$$
$$+C\|\nabla\partial_t b\|_{L^1(0,T;L^\infty(\Omega))}\left\||\nabla(b-b_M^-)|^3+|\epsilon(b-b_M^-)|+1\right\|_{L^\infty(0,T;L^1(\Omega))}$$
$$\leq\mathcal{E}_\varepsilon(q_\varepsilon^0)+C\int_{-\tau}^{d_M(t)}\|\nabla\partial_t b(s+\tau)\|_{L^\infty(\Omega)}\mathcal{E}_\varepsilon(q_M(s))\,\mathrm{d}s+C$$
$$\leq\mathcal{E}_\varepsilon(q_\varepsilon^0)+C\int_{-\tau}^{t}\|\nabla\partial_t b(s+\tau)\|_{L^\infty(\Omega)}\gamma(s)\,\mathrm{d}s+C.$$

Gronwall's inequality shows for all $t\in[0,T]$:

$$\gamma(t)\leq C+\mathcal{E}_\varepsilon(q_\varepsilon^0)$$
$$+C\int_{-\tau}^t(C+\mathcal{E}_\varepsilon(q_\varepsilon^0))\|\nabla\partial_t b(s+\tau)\|_{L^\infty(\Omega)}\exp\left(\int_s^t\|\nabla\partial_t b(l+\tau)\|_{L^\infty(\Omega)}\,\mathrm{d}l\right)\mathrm{d}s$$
$$\leq C(\mathcal{E}_\varepsilon(q_\varepsilon^0)+1).$$

$\square$

Corollary 4.3.4 (A-priori estimates) *There exists a constant $C>0$ independent of M such that*

(i) $\|u_M\|_{L^\infty(0,T;W^{1,4}(\Omega;\mathbb{R}^n))}\leq C,$ (iv) $\|\partial_t\widehat{c}_M\|_{L^2(\Omega_T)}\leq C,$

(ii) $\|c_M\|_{L^\infty(0,T;H^1(\Omega))}\leq C,$ (v) $\|\partial_t\widehat{z}_M\|_{L^2(\Omega_T)}\leq C,$

(iii) $\|z_M\|_{L^\infty(0,T;W^{1,p}(\Omega))}\leq C,$ (vi) $\|\mu_M\|_{L^2(0,T;H^1(\Omega))}\leq C$

for all $M\in\mathbb{N}$.

Proof. We use Lemma 4.3.3. The boundedness of $\nabla(u_M(t) - b_M(t))$ in $L^4(\Omega; \mathbb{R}^n)$ with respect to $M \in \mathbb{N}$ and $t \in [0, T]$ and $u_M(t) - b_M(t) \in H^1_{\Gamma_D}(\Omega; \mathbb{R}^n)$ yield (i) by Poincaré's inequality. The boundedness of $\nabla c_M(t)$ in $L^2(\Omega)$ and mass conservation imply (ii) by Poincaré's inequality. The boundedness of $\nabla z_M(t)$ in $L^p(\Omega)$ and $0 \leq z_M(t) \leq 1$ in Ω for all M and all t show (iii). The properties (iv) and (v) follow immediately from Lemma 4.3.3. The boundedness of $\nabla \mu_M$ in $L^2(\Omega_T)$ and $\int_\Omega \mu_M(t) \, dx$ with respect to M and t show (vi) by Poincaré's inequality. Indeed, $\int_\Omega \mu_M(t) \, dx$ is bounded with respect to M and t because of (4.27) and (4.26) tested with $\zeta \equiv 1$. $\qquad \square$

Due to the a-priori estimates, we can select weakly (weakly-$\star$) convergent subsequences (see Lemma 4.3.5). Furthermore, exploiting the Euler-Lagrange equations of the approximate solutions, we will even attain strong convergence properties (see Lemma 4.3.6 and Lemma 4.3.8).

Lemma 4.3.5 (Weak convergence of the approximate solutions) *There exists a subsequence $\{M_k\}$ and functions (c, u, z, μ) satisfying Definition 4.2.3 (i) such that the following properties are satisfied:*

(i) $z_{M_k}, z^-_{M_k}, \widehat{z}_{M_k} \overset{\star}{\rightharpoonup} z$ *in* $L^\infty(0, T; W^{1,p}(\Omega))$,
   $z_{M_k}(t), z^-_{M_k}(t), \widehat{z}_{M_k}(t) \rightharpoonup z(t)$ *in* $W^{1,p}(\Omega)$ *for a.e.* t,
   $z_{M_k}, z^-_{M_k}, \widehat{z}_{M_k} \to z$ *a.e. in* Ω_T,
   $\widehat{z}_{M_k} \rightharpoonup z$ *in* $H^1(0, T; L^2(\Omega))$,

(ii) $u_{M_k} \overset{\star}{\rightharpoonup} u$ *in* $L^\infty(0, T; W^{1,4}(\Omega; \mathbb{R}^n))$,

(iii) $c_{M_k}, c^-_{M_k}, \widehat{c}_{M_k} \overset{\star}{\rightharpoonup} c$ *in* $L^\infty(0, T; H^1(\Omega))$,
   $c_{M_k}(t), c^-_{M_k}(t), \widehat{c}_{M_k}(t) \rightharpoonup c(t)$ *in* $H^1(\Omega)$ *for a.e.* t,
   $c_{M_k}, c^-_{M_k}, \widehat{c}_{M_k} \to c$ *a.e. in* Ω_T,
   $\widehat{c}_{M_k} \rightharpoonup c$ *in* $H^1(0, T; L^2(\Omega))$,

(iv) $\mu_{M_k} \rightharpoonup \mu$ *in* $L^2(0, T; H^1(\Omega))$

as $k \to \infty$.

Proof. To simplify notation, we omit the index k in the proof.

(iii) Since $\widehat{c}_M$ is bounded in $L^2(0, T; H^1(\Omega))$ and $\partial_t \widehat{c}_M$ is bounded in $L^2(\Omega_T)$, we obtain $\widehat{c}_M \to \widehat{c}$ in $L^2(\Omega_T)$ as $M \to \infty$ for a subsequence by a compactness result from J. P. Aubin and J. L. Lions (see Theorem 2.3.9 (i)). Therefore, we can extract a subsequence such that $\widehat{c}_M(t) \to \widehat{c}(t)$ in $L^2(\Omega)$ for a.e. $t \in (0, T)$ and $\widehat{c}_M \to \widehat{c}$ a.e. in Ω_T. We denote this subsequence also with $\{\widehat{c}_M\}$. The boundedness of $\{\widehat{c}_M(t)\}_{M \in \mathbb{N}}$ in $H^1(\Omega)$ even shows $\widehat{c}_M(t) \rightharpoonup \widehat{c}(t)$ in $H^1(\Omega)$ for a.e. $t \in (0, T)$. In addition, the boundedness of $\{\widehat{c}_M\}$ in $L^\infty(0, T; H^1(\Omega))$ shows $\widehat{c}_M \overset{\star}{\rightharpoonup} \widehat{c}$ in $L^\infty(0, T; H^1(\Omega))$. Furthermore, we obtain from the boundedness of $\{\partial_t \widehat{c}_M\}$ in $L^2(\Omega_T)$ for every $t \in [0, T]$:

$$\|c_M(t) - \widehat{c}_M(t)\|_{L^1(\Omega)} = \|\widehat{c}_M(d_M(t)) - \widehat{c}_M(t)\|_{L^1(\Omega)}$$

$$\leq \int_t^{d_M(t)} \|\partial_t \widehat{c}_M(s)\|_{L^1(\Omega)} \, \mathrm{d}s$$

$$\leq C(d_M(t) - t)^{1/2} \|\partial_t \widehat{c}_M\|_{L^2(\Omega_T)} \to 0 \text{ as } M \to \infty.$$

Lebesgue's convergence theorem yields $\|c_M - \widehat{c}_M\|_{L^1(\Omega_T)} \to 0$ as $M \to \infty$. Analogously, we obtain $\|c_M - c_M^-\|_{L^1(\Omega_T)} \to 0$ as $M \to \infty$. Thus, the convergence properties for $\widehat{c}_M$ also holds for c_M and c_M^- with the same limit $c = c^- = \widehat{c}$ a.e. . The boundedness of $\{\widehat{c}_M\}$ in $H^1(0, T; L^2(\Omega))$ shows $\widehat{c}_M \rightharpoonup c$ in $H^1(0, T; L^2(\Omega))$ for a subsequence.

(i) We obtain the convergence properties for $\{z_M\}$, $\{z_M^-\}$ and $\{\widehat{z}_M\}$ with the same argumentation as in (iii). Note that the limit function also satisfies $\partial_t z \leq 0$ a.e. in Ω_T.

Assume the contrary. Then, there exists a measurable set $G \subseteq \Omega_T$ with $\mathcal{L}^{n+1}(G) > 0$ and $\partial_t z > 0$ a.e. in G. By $\partial_t \widehat{z}_M \rightharpoonup \partial_t z$ in $L^2(\Omega_T)$ and $\partial_t \widehat{z}_M \leq 0$ a.e. in Ω_T for all $M \in \mathbb{N}$, we obtain as $M \to \infty$:

$$0 \geq \int_G \partial_t \widehat{z}_M \, \mathrm{d}x \, \mathrm{d}t = \int_{\Omega_T} (\partial_t \widehat{z}_M) \chi_G \, \mathrm{d}x \, \mathrm{d}t \to \int_{\Omega_T} (\partial_t z) \chi_G \, \mathrm{d}x \, \mathrm{d}t = \int_G \partial_t z \, \mathrm{d}x \, \mathrm{d}t.$$

Therefore, $\int_G \partial_t z \, \mathrm{d}x \, \mathrm{d}t \leq 0$ which is a contradiction.

(ii) This property follows from the boundedness of $\{u_M\}$ in $L^\infty(0, T; H^1(\Omega; \mathbb{R}^n))$.

(iv) This property follows from the boundedness of $\{\mu_M\}$ in $L^2(0, T; H^1(\Omega))$. $\qquad\square$

In the sequel, we take advantage from the elementary inequality (x, y are elements of an inner product space X with scalar product $\langle \cdot, \cdot \rangle$)

$$C_{\mathrm{uc}} \|x - y\|^q \leq \langle (\|x\|^{q-2} x - \|y\|^{q-2} y), x - y \rangle \tag{4.37}$$

for a constant $C_{\mathrm{uc}} > 0$ depending on X and $q \geq 2$. To see this, (4.37) is equivalent to

$$C_{\mathrm{uc}} \leq \langle b, \|a + b\|^{q-2}(a + b) - \|a\|^{q-2} a \rangle \text{ for all } a, b \in X, \|b\| = 1$$

by introducing the variables $a := x/\|x - y\|$ and $b := (x - y)/\|x - y\|$ for $x \neq y$. This is equivalent to

$$C_{\mathrm{uc}} \leq \|a + b\|^{q-2} + \langle b, a \rangle (\|a + b\|^{q-2} - \|a\|^{q-2}) \text{ for all } a, b \in X, \|b\| = 1. \tag{4.38}$$

Now the equivalence $\|a + b\| \leq \|a\| \Leftrightarrow \langle a, b \rangle \leq -\frac{1}{2} \|b\|^2$ as well as $\|b\| = 1$ give the estimate ($\eta > 0$ constant):

$$\|a + b\|^{q-2} + \langle b, a \rangle (\|a + b\|^{q-2} - \|a\|^{q-2}) \geq \|a + b\|^{q-2} + \frac{1}{2} \|b\|^2 (\|a\|^{q-2} - \|a + b\|^{q-2})$$

$$= \frac{1}{2} \|a + b\|^{q-2} + \frac{1}{2} \|a\|^{q-2}$$

$$\geq \eta(\|a + b\| + \|a\|)^{q-2}$$

$$\geq \eta(\|b\| - \|a\| + \|a\|)^{q-2}$$

$$= \eta > 0$$

Therefore, (4.38) follows.

Lemma 4.3.6 *There exists a subsequence $\{M_k\}$ such that*

$$u_{M_k}, u_{M_k}^- \to u \ \text{in} \ L^4(0, T; W^{1,4}(\Omega; \mathbb{R}^n)) \ \text{as} \ k \to \infty.$$

Proof. We omit the index k in the proof.
Applying (4.2a), taking inequality (4.37) for $q = 4$ into account and considering (4.28) with the test function $\zeta = u_M(t) - u(t) - b_M(t) + b(t)$, we get

$$\eta \|\epsilon(u_M) - \epsilon(u)\|^2_{L^2(\Omega_T; \mathbb{R}^{n \times n})} + \varepsilon C_{\text{uc}} \|\nabla u_M - \nabla u\|^4_{L^4(\Omega_T; \mathbb{R}^{n \times n})}$$

$$\leq \int_{\Omega_T} (\partial_e W^{\text{el}}(c_M, \epsilon(u_M), z_M) - \partial_e W^{\text{el}}(c_M, \epsilon(u), z_M)) : (\epsilon(u_M) - \epsilon(u)) \, dx \, dt$$

$$+ \varepsilon \int_{\Omega_T} (|\nabla u_M|^2 \nabla u_M - |\nabla u|^2 \nabla u) : (\nabla u_M - \nabla u) \, dx \, dt$$

$$= \underbrace{\int_{\Omega_T} (\partial_e W^{\text{el}}(c_M, \epsilon(u_M), z_M) : \epsilon(\zeta) + \varepsilon |\nabla u_M|^2 \nabla u_M : \nabla \zeta) \, dx \, dt}_{=0 \ \text{by (4.28)}}$$

$$+ \underbrace{\int_{\Omega_T} \partial_e W^{\text{el}}(c_M, \epsilon(u_M), z_M) : (\epsilon(b_M) - \epsilon(b)) \, dx \, dt}_{(\star)}$$

$$+ \varepsilon \underbrace{\int_{\Omega_T} |\nabla u_M|^2 \nabla u_M : (\nabla b_M - \nabla b) \, dx \, dt}_{(\star\star)}$$

$$- \underbrace{\int_{\Omega_T} (\partial_e W^{\text{el}}(c_M, \epsilon(u), z_M) : (\epsilon(u_M) - \epsilon(u)) \, dx \, dt}_{(\star\star\star)}$$

$$- \varepsilon \underbrace{\int_{\Omega_T} |\nabla u|^2 \nabla u : (\nabla u_M - \nabla u) \, dx \, dt}_{(\star\star\star\star)}. \tag{4.39}$$

Since $\partial_e W^{\text{el}}(c_M, \epsilon(u_M), z_M)$ is bounded in $L^2(\Omega_T; \mathbb{R}^{n \times n})$ (by (4.3) and Corollary 4.3.4) as well as $\epsilon(b_M) \to e(b)$ in $L^2(\Omega_T; \mathbb{R}^{n \times n})$, we obtain $(\star) \to 0$ as $M \to \infty$. The boundedness of $|\nabla u_M|^2 \nabla u_M$ in $L^{4/3}(\Omega_T; \mathbb{R}^{n \times n})$ by Corollary 4.3.4 and $\nabla b_M \to \nabla b$ in $L^4(\Omega_T; \mathbb{R}^{n \times n})$ lead to $(\star\star) \to 0$. We also have $\partial_e W^{\text{el}}(c_M, \epsilon(u), z_M) \to \partial_e W^{\text{el}}(c, \epsilon(u), z)$ in $L^2(\Omega_T; \mathbb{R}^{n \times n})$ by (4.3) and Lebesgue's generalized convergence theorem.

Furthermore, $\epsilon(u_M) \rightharpoonup \epsilon(u)$ in $L^2(\Omega_T; \mathbb{R}^n \times \mathbb{R}^n)$ by Lemma 4.3.5. This gives $(\star\star\star) \to 0$. Since $\nabla u_M \rightharpoonup \nabla u$ in $L^4(\Omega_T; \mathbb{R}^{n \times n})$ by Lemma 4.3.5, we obtain $(\star\star\star\star) \to 0$. Therefore, (4.39) implies $\epsilon(u_M) \to \epsilon(u)$ in $L^2(\Omega_T; \mathbb{R}^{n \times n})$ and $\nabla u_M \to \nabla u$ in $L^4(\Omega_T; \mathbb{R}^{n \times n})$ as $M \to \infty$. Poincaré's inequality finally shows $u_M \to u$ in $L^4(0, T; W^{1,4}(\Omega; \mathbb{R}^n))$. Now, we choose a subsequence such that $u_M(t) \to u(t)$ in $W^{1,4}(\Omega; \mathbb{R}^n)$ for a.e. $t \in (0, T)$ and $u_M \to u$ a.e. in Ω_T. We also denote this subsequence with $\{u_M\}$.

Analogously, we obtain a $u^- \in L^4(0,T;W^{1,4}(\Omega))$ satisfying

$$u_M^- \to u^- \text{ in } L^4(0,T;W^{1,4}(\Omega)), \tag{4.40a}$$

$$u_M^- \to u^- \text{ a.e. in } \Omega_T. \tag{4.40b}$$

We will show $u = u^-$ a.e. in Ω_T. Consider the variational property (4.28) for $q_M(t)$ and for $q_M^-(t)$:

$$0 = \int_{\Omega_T} \left(\partial_e W^{\text{el}}(c_M, \epsilon(u_M), z_M) : \epsilon(\zeta) + \varepsilon |\nabla u_M|^2 \nabla u_M : \nabla \zeta \right) dx\, dt, \tag{4.41a}$$

$$0 = \int_{\Omega_T} \left(\partial_e W^{\text{el}}(c_M^-, \epsilon(u_M^-), z_M^-) : \epsilon(\zeta) + \varepsilon |\nabla u_M^-|^2 \nabla u_M^- : \nabla \zeta \right) dx\, dt. \tag{4.41b}$$

We choose the test function $\zeta(t) = u_M(t) - u_M^-(t) - b_M(t) + b_M^-(t) \in W^{1,4}_{\Gamma_D}(\Omega)$. An estimate similar to (4.39) gives

$$\eta \|\epsilon(u_M) - \epsilon(u_M^-)\|^2_{L^2(\Omega_T)} + \varepsilon C^{-1}_{\text{ineq}} \|\nabla u_M - \nabla u_M^-\|^4_{L^4(\Omega_T)}$$

$$\leq \int_{\Omega_T} \left(\partial_e W^{\text{el}}(c_M, \epsilon(u_M), z_M) - \partial_e W^{\text{el}}(c_M, \epsilon(u_M^-), z_M) \right) : (\epsilon(u_M) - \epsilon(u_M^-))\, dx\, dt$$

$$+ \varepsilon \int_{\Omega_T} \left(|\nabla u_M|^2 \nabla u_M - |\nabla u_M^-|^2 \nabla u_M^- \right) : (\nabla u_M - \nabla u_M^-)\, dx\, dt$$

$$= \underbrace{\int_{\Omega_T} \left(\partial_e W^{\text{el}}(c_M, \epsilon(u_M), z_M) : \epsilon(\zeta) + \varepsilon |\nabla u_M|^2 \nabla u_M : \nabla \zeta \right) dx\, dt}_{=0 \text{ by (4.41a)}}$$

$$- \underbrace{\int_{\Omega_T} \left(\partial_e W^{\text{el}}(c_M^-, \epsilon(u_M^-), z_M^-) : \epsilon(\zeta) + \varepsilon |\nabla u_M^-|^2 \nabla u_M^- : \nabla \zeta \right) dx\, dt}_{=0 \text{ by (4.41b)}}$$

$$+ \int_{\Omega_T} \left(\partial_e W^{\text{el}}(c_M^-, \epsilon(u_M^-), z_M^-) - \partial_e W^{\text{el}}(c_M, \epsilon(u_M^-), z_M) \right) : (\epsilon(u_M) - \epsilon(u_M^-))\, dx\, dt$$

$$+ \int_{\Omega_T} \left(\partial_e W^{\text{el}}(c_M, \epsilon(u_M), z_M) - \partial_e W^{\text{el}}(c_M^-, \epsilon(u_M^-), z_M^-) \right) : (\epsilon(b_M) - \epsilon(b_M^-))\, dx\, dt$$

$$+ \varepsilon \int_{\Omega_T} \left(|\nabla u_M|^2 \nabla u_M - |\nabla u_M^-|^2 \nabla u_M^- \right) : (\nabla b_M - \nabla b_M^-)\, dx\, dt. \tag{4.42}$$

We need to prove that

$$\partial_e W^{\text{el}}(c_M^-, \epsilon(u_M^-), z_M^-) - \partial_e W^{\text{el}}(c_M, \epsilon(u_M^-), z_M) \to 0 \text{ in } L^2(\Omega_T; \mathbb{R}^{n \times n}). \tag{4.43}$$

- In fact, we already know from Lemma 4.3.5 and from (4.40b) that $c_M^-, c_M \to c$, $u_M^- \to u^-$ and $z_M^-, z_M \to z$ as $M \to \infty$ a.e. in Ω_T. It follows from the continuity of the function $\partial_e W^{\text{el}}$ that

$$\partial_e W^{\text{el}}(c_M^-, \epsilon(u_M^-), z_M^-) - \partial_e W^{\text{el}}(c_M, \epsilon(u_M^-), z_M)$$

$$\to \partial_e W^{\mathrm{el}}(c, \epsilon(u^-), z) - \partial_e W^{\mathrm{el}}(c, \epsilon(u^-), z) = 0 \text{ a.e. in } \Omega_T.$$

The growth condition (4.3) shows

$$\left| \partial_e W^{\mathrm{el}}(c_M^-, \epsilon(u_M^-), z_M^-) - \partial_e W^{\mathrm{el}}(c_M, \epsilon(u_M^-), z_M) \right|^2 \le C(|c_M^-|^4 + |c_M|^4 + |\epsilon(u_M^-)|^2 + 1).$$

From Lemma 4.3.5 and (4.40a), we can conclude

$$|c_M^-|^4 + |c_M|^4 + |\epsilon(u_M^-)|^2 \to |c^-|^4 + |c|^4 + |\epsilon(u^-)|^2 \text{ in } L^1(\Omega_T).$$

Lebesgue's generalized convergence theorem yields (4.43).

We also know $\epsilon(b_M) - \epsilon(b_M^-) \to 0$ in $L^2(\Omega_T; \mathbb{R}^{n \times n})$ and $\nabla b_M - \nabla b_M^- \to 0$ in $L^4(\Omega_T; \mathbb{R}^{n \times n})$. Hence, each term on the right hand side in (4.42) converges to 0 as $M \to \infty$ $\square$

Lemma 4.3.7 *There exists a subsequence* $\{M_k\}$ *such that*

$$c_{M_k}, c_{M_k}^-, \widehat{c}_{M_k} \to c \text{ in } L^2(0, T; H^1(\Omega)) \text{ as } k \to \infty.$$

Proof. We omit the index k in the proof.
Lemma 4.3.5 implies $c_M(t) \to c(t)$ in $L^{2^*/2+1}(\Omega)$ for a.e. $t \in (0, T)$. Using Corollary 4.3.4 and Lebesgue's convergence theorem, we get $c_M \to c$ in $L^{2^*/2+1}(\Omega_T)$. Next, we test (4.27) with $\zeta = c_M(t)$, integrate from $t = 0$ to $t = T$ and obtain

$$\int_{\Omega_T} |\nabla c_M|^2 \, \mathrm{d}x \, \mathrm{d}t = - \int_{\Omega_T} \left(\partial_c W^{\mathrm{ch}}(c_M) + \partial_c W^{\mathrm{el}}(c_M, \epsilon(u_M), z_M) + \varepsilon(\partial_t \widehat{c}_M) - \mu_M \right) c_M \, \mathrm{d}x \, \mathrm{d}t.$$

We know the following:

- $\partial_t \widehat{c}_M \rightharpoonup \partial_t c$ and $\mu_M \rightharpoonup \mu$ in $L^2(\Omega_T)$ due to Lemma 4.3.5. Because of $c_M \to c$ in $L^2(\Omega_T)$, we get

$$\int_{\Omega_T} \left(\varepsilon(\partial_t \widehat{c}_M) - \mu_M \right) c_M \, \mathrm{d}x \, \mathrm{d}t \to \int_{\Omega_T} \left(\varepsilon(\partial_t c - \mu) \right) c \, \mathrm{d}x \, \mathrm{d}t.$$

- Due to Lemma 4.3.5, we obtain the pointwise convergence:

$$\partial_c W^{\mathrm{ch}}(c_M) c_M \to \partial_c W^{\mathrm{ch}}(c) c \text{ a.e. in } \Omega_T.$$

Growth assumption (4.2g) shows

$$\left| \partial_c W^{\mathrm{ch}}(c_M) c_M \right| \le C(|c_M|^{2^*/2+1} + 1).$$

Lemma 4.3.5 and the compact embedding $H^1(\Omega) \hookrightarrow L^{2^*/2+1}(\Omega)$ yield the convergence $|c_M|^{2^*/2+1} \to |c|^{2^*/2+1}$ in $L^1(\Omega)$. Thus, Lebesgue's generalized convergence theorem shows

$$\int_{\Omega_T} \partial_c W^{\mathrm{ch}}(c_M) c_M \, \mathrm{d}x \, \mathrm{d}t \to \int_{\Omega_T} \partial_c W^{\mathrm{ch}}(c) c \, \mathrm{d}x \, \mathrm{d}t.$$

- Due to Lemma 4.3.5 and Lemma 4.3.6, there exists a subsequence such that the pointwise convergence holds:

$$\partial_c W^{\text{el}}(c_M, \epsilon(u_M), z_M) \to \partial_c W^{\text{el}}(c, \epsilon(u), z) \text{ a.e. in } \Omega_T.$$

Beyond that, we obtain due to (4.2d) the estimate

$$\left|\partial_c W^{\text{el}}(c_M, \epsilon(u_M), z_M)\right|^2 \leq C(|c_M|^4 + |\epsilon(u_M)|^2 + 1).$$

Since the right hand side converges strongly in $L^1(\Omega_T)$, Lebesgue's generalized convergence theorem yields

$$\partial_c W^{\text{el}}(c_M, \epsilon(u_M), z_M) \to \partial_c W^{\text{el}}(c, \epsilon(u), z) \text{ in } L^2(\Omega_T).$$

Taking also $c_M \to c$ in $L^2(\Omega_T)$ into account, we get

$$\int_{\Omega_T} \partial_c W^{\text{el}}(c_M, \epsilon(u_M), z_M) c_M \, dx \, dt \to \int_{\Omega_T} \partial_c W^{\text{el}}(c, \epsilon(u), z) c \, dx \, dt.$$

Summing up, we obtain

$$\int_{\Omega_T} |\nabla c_M|^2 \, dx \, dt \to - \int_{\Omega_T} \left(\partial_c W^{\text{ch}}(c) + \partial_c W^{\text{el}}(c, \epsilon(u), z) + \varepsilon(\partial_t c) - \mu\right) c \, dx \, dt$$

as $M \to \infty$. On the other hand, we test (4.27) with $c(t)$ and integrate from $t = 0$ to $t = T$. Note that $c \in L^{2^*}(\Omega_T)$ and $\partial_c W^{\text{ch}}(c_M) \to \partial_c W^{\text{ch}}(c)$ in $L^{2^*/(2^*-1)}(\Omega_T)$ as $M \to \infty$ by Lebesgue's generalized convergence theorem. Hence, we derive for $M \to \infty$:

$$\int_{\Omega_T} |\nabla c|^2 \, dx \, dt = - \int_{\Omega_T} \left(\partial_c W^{\text{ch}}(c) c + \partial_c W^{\text{el}}(c, \epsilon(u), z) c + \varepsilon(\partial_t c) c - \mu c\right) dx \, dt.$$

Therefore, $c_M \to c$ in $L^2(0, T; H^1(\Omega))$ as $M \to \infty$. The convergence $\|c_M\|_{L^2(0,T;H^1(\Omega))} \to \|c\|_{L^2(0,T;H^1(\Omega))}$ implies $\|c_M^-\|_{L^2(0,T;H^1(\Omega))} \to \|c\|_{L^2(0,T;H^1(\Omega))}$. We also have $c_M^- \rightharpoonup c$ in $L^2(0, T; H^1(\Omega))$ (by Lemma 4.3.5 (ii)) and consequently $c_M^- \to c$ and $\widehat{c}_M \to c$ in $L^2(0, T; H^1(\Omega))$ as $M \to \infty$. $\qquad\square$

Note that in connection with Corollary 4.3.4 we even get for each $q \geq 1$

$$c_M, c_M^-, \widehat{c}_M \to c \text{ in } L^q(0, T; H^1(\Omega))$$

for a subsequence as $M \to \infty$.

Lemma 4.3.8 *There exists a subsequence $\{M_k\}$ such that*

$$z_{M_k}, z_{M_k}^-, \widehat{z}_{M_k} \to z \text{ in } L^p(0, T; W^{1,p}(\Omega)) \text{ as } k \to \infty.$$

Proof. To simplify notation, we omit the index k in the proof.

Applying Lemma 2.3.18 with $f = \zeta = z$ and $f_M = z_M^-$ gives a sequence of approximations $\{\zeta_M\}_{M \in \mathbb{N}} \subseteq L^p(0, T; W_+^{1,p}(\Omega)) \cap L^\infty(\Omega_T)$ with the properties (note that we have $z_M^-(t) \rightharpoonup z(t)$ in $W^{1,p}(\Omega)$ for a.e. $t \in (0, T)$ by Lemma 4.3.5):

$$\zeta_M \to z \text{ in } L^p(0, T; W^{1,p}(\Omega)) \text{ as } M \to \infty \tag{4.44a}$$

$$0 \le \zeta_M \le z_M^- \text{ a.e. in } \Omega_T \text{ for all } M \in \mathbb{N}. \tag{4.44b}$$

We test (4.29) with $\zeta = \zeta_M(t) - z_M(t)$ for $\nu = 1$ (possible due to (4.44b)), integrate from $t = 0$ to $t = T$ and use (4.37) to obtain the following estimate:

$$
\begin{aligned}
C_{\text{uc}} &\int_{\Omega_T} |\nabla z_M - \nabla z|^p \, dx \, dt \\
&\le \int_{\Omega_T} (|\nabla z_M|^{p-2} \nabla z_M - |\nabla z|^{p-2} \nabla z) \cdot \nabla(z_M - z) \, dx \, dt \\
&= \int_{\Omega_T} |\nabla z_M|^{p-2} \nabla z_M \cdot \nabla(z_M - \zeta_M) \, dx \, dt \\
&\quad + \int_{\Omega_T} \left(|\nabla z_M|^{p-2} \nabla z_M \cdot \nabla(\zeta_M - z) - |\nabla z|^{p-2} \nabla z \cdot \nabla(z_M - z)\right) dx \, dt \\
&\le \int_{\Omega_T} (\partial_z W^{\text{el}}(c_M, \epsilon(u_M), z_M) - \alpha + \beta \partial_t \widehat{z}_M)(\zeta_M - z_M) \, dx \, dt \\
&\quad + \int_{\Omega_T} \left(|\nabla z_M|^{p-2} \nabla z_M \cdot \nabla(\zeta_M - z) - |\nabla z|^{p-2} \nabla z \cdot \nabla(z_M - z)\right) dx \, dt \\
&\le \underbrace{\|\partial_z W^{\text{el}}(c_M, \epsilon(u_M), z_M) - \alpha + \beta \partial_t \widehat{z}_M\|_{L^2(\Omega_T)}}_{\text{bounded by (4.2f) and Cor. 4.3.4}} \|\zeta_M - z_M\|_{L^2(\Omega_T)} \\
&\quad + \underbrace{\|\nabla z_M\|_{L^p(\Omega_T)}^{p-1}}_{\text{bounded by Cor. 4.3.4}} \|\nabla \zeta_M - \nabla z\|_{L^p(\Omega_T)} - \int_{\Omega_T} |\nabla z|^{p-2} \nabla z \cdot \nabla(z_M - z) \, dx \, dt.
\end{aligned}
$$

Observe that $\nabla \zeta_M - \nabla z \to 0$ in $L^p(\Omega_T; \mathbb{R}^n)$ and $\zeta_M - z_M \to 0$ in $L^2(\Omega_T)$ (by property (4.44a) and by Lemma 4.3.5) as well as $\nabla z_M - \nabla z \rightharpoonup 0$ in $L^p(\Omega_T; \mathbb{R}^n)$ by Lemma 4.3.5. Using these properties, each term on the right hand side converges to 0 as $M \to \infty$.

We also obtain $\|z_M^-\|_{L^p(0,T;W^{1,p}(\Omega))} \to \|z\|_{L^p(0,T;W^{1,p}(\Omega))}$ from $\|z_M\|_{L^p(0,T;W^{1,p}(\Omega))} \to \|z\|_{L^p(0,T;W^{1,p}(\Omega))}$. Because of $z_M^- \rightharpoonup z$ in $L^p(0, T; W^{1,p}(\Omega))$ (by Lemma 4.3.5 (i)), we even have $z_M^- \to z$ and $\widehat{z}_M \to z$ in $L^p(0, T; W^{1,p}(\Omega))$ as $M \to \infty$. $\qquad\square$

In conclusion, Corollary 4.3.4, Lemma 4.3.5, Lemma 4.3.6, Lemma 4.3.7 and Lemma 4.3.8 imply the following convergence properties:

Corollary 4.3.9 *There exists subsequence $\{M_k\}$ and functions (c, u, z, μ) satisfying Definition 4.2.3 (i) such that the following properties are satisfied as $k \to \infty$:*

(i) $z_{M_k}, z_{M_k}^-, \widehat{z}_{M_k} \to z$ in $L^p(0, T; W^{1,p}(\Omega))$,

$z_{M_k}(t), z_{M_k}^-(t), \widehat{z}_{M_k}(t) \to z(t)$ in $W^{1,p}(\Omega)$ for a.e. t,

$z_{M_k}, z_{M_k}^-, \widehat{z}_{M_k} \to z$ a.e. in Ω_T,

$\widehat{z}_{M_k} \rightharpoonup z$ in $H^1(0, T; L^2(\Omega))$,

(ii) $c_{M_k}, c_{M_k}^-, \widehat{c}_{M_k} \to c$ in $L^{2^*}(0, T; H^1(\Omega))$,

$c_{M_k}(t), c_{M_k}^-(t), \widehat{c}_{M_k}(t) \to c(t)$ in $H^1(\Omega)$ for a.e. t,

$c_{M_k}, c_{M_k}^-, \widehat{c}_{M_k} \to c$ a.e. in Ω_T,

$\widehat{c}_{M_k} \rightharpoonup c$ in $H^1(0, T; L^2(\Omega))$,

(iii) $u_{M_k}, u_{M_k}^- \to u$ in $L^4(0, T; W^{1,4}(\Omega; \mathbb{R}^n))$,

$u_{M_k}(t), u_{M_k}^-(t) \to u(t)$ in $W^{1,4}(\Omega; \mathbb{R}^n)$ for a.e. t,

$u_{M_k}, u_{M_k}^- \to u$ a.e. in Ω_T,

(iv) $\mu_{M_k} \rightharpoonup \mu$ in $L^2(0, T; H^1(\Omega))$,

In the following, we omit the subscript k.

The above convergence properties allow us to establish an energy estimate, which is in an asymptotic sense stronger than the one in Lemma 4.3.2 (v). We emphasize that (4.30) has in comparison with (4.45) no factor $1/2$ in front of the terms $\beta|\partial_t \widehat{z}_M|^2$, $\varepsilon|\partial_t \widehat{c}_M|^2$ and $|\nabla \mu_M|^2$.

Lemma 4.3.10 (Precise energy inequality) *For every* $0 \le t_1 < t_2 \le T$:

$$\mathcal{E}_\varepsilon(q_M(t_2)) + \int_{d_M^-(t_1)}^{d_M(t_2)} \int_\Omega \left(-\alpha \partial_t \widehat{z}_M + \beta|\partial_t \widehat{z}_M|^2 + \varepsilon|\partial_t \widehat{c}_M|^2 + |\nabla \mu_M|^2 \right) dx \, ds - \mathcal{E}_\varepsilon(q_M^-(t_1))$$

$$\le \int_{d_M^-(t_1)}^{d_M(t_2)} \int_\Omega \partial_e W^{el}(c_M^-, \epsilon(u_M^- + b - b_M^-), z_M^-) : \epsilon(\partial_t b) \, dx \, ds$$

$$+ \varepsilon \int_{d_M^-(t_1)}^{d_M(t_2)} \int_\Omega |\nabla u_M^- + \nabla b - \nabla b_M^-|^2 \nabla(u_M^- + b - b_M^-) : \nabla \partial_t b \, dx \, ds + \kappa_M$$

$$\tag{4.45}$$

with $\kappa_M \to 0$ *as* $M \to \infty$.

Proof. We know $\mathbb{E}_M^m(q_M^m) \le \mathbb{E}_M^m(u_M^{m-1} + b_M^m - b_M^{m-1}, c_M^m, z_M^m)$. The regularity properties of the functions b, $\widehat{c}_M$ and $\widehat{z}_M$ ensure that the chain rule can be applied and the following integral terms are well defined:

$$\mathcal{E}_\varepsilon(c_M^m, u_M^m, z_M^m)$$
$$\le \mathcal{E}_\varepsilon(c_M^m, u_M^{m-1} + b_M^m - b_M^{m-1}, z_M^m)$$
$$= \mathcal{E}_\varepsilon(c_M^{m-1}, u_M^{m-1}, z_M^{m-1})$$
$$+ \mathcal{E}_\varepsilon(c_M^{m-1}, u_M^{m-1} + b_M^m - b_M^{m-1}, z_M^{m-1}) - \mathcal{E}_\varepsilon(c_M^{m-1}, u_M^{m-1}, z_M^{m-1})$$
$$+ \mathcal{E}_\varepsilon(c_M^m, u_M^{m-1} + b_M^m - b_M^{m-1}, z_M^{m-1}) - \mathcal{E}_\varepsilon(c_M^{m-1}, u_M^{m-1} + b_M^m - b_M^{m-1}, z_M^{m-1})$$

$$+ \mathcal{E}_\varepsilon(c_M^m, u_M^{m-1} + b_M^m - b_M^{m-1}, z_M^m) - \mathcal{E}_\varepsilon(c_M^m, u_M^{m-1} + b_M^m - b_M^{m-1}, z_M^{m-1})$$

$$= \mathcal{E}_\varepsilon(c_M^{m-1}, u_M^{m-1}, z_M^{m-1})$$

$$+ \int_{(m-1)\tau}^{m\tau} \langle d_u \widetilde{\mathcal{E}}_\varepsilon(c_M^{m-1}, u_M^{m-1} + b(s) - b_M^{m-1}, z_M^{m-1}), \partial_t b(s)\rangle_{(H^1)^* \times H^1} \, ds$$

$$+ \int_{(m-1)\tau}^{m\tau} \langle d_c \widetilde{\mathcal{E}}_\varepsilon(\widehat{c}_M(s), u_M^{m-1} + b_M^m - b_M^{m-1}, z_M^{m-1}), \partial_t \widehat{c}_M(s)\rangle_{(H^1)^* \times H^1} \, ds$$

$$+ \int_{(m-1)\tau}^{m\tau} \langle d_z \widetilde{\mathcal{E}}_\varepsilon(c_M^m, u_M^{m-1} + b_M^m - b_M^{m-1}, \widehat{z}_M(s)), \partial_t \widehat{z}_M(s)\rangle_{(W^{1,p})^* \times W^{1,p}} \, ds.$$

Summing from $m = \frac{d_M^-(t_1)}{\tau} + 1$ to $\frac{d_M(t_2)}{\tau}$ yields

$$\mathcal{E}_\varepsilon(q_M(t_2)) - \mathcal{E}_\varepsilon(q_M^-(t_1))$$

$$\leq \varepsilon \int_{d_M^-(t_1)}^{d_M(t_2)} \int_\Omega |\nabla(u_M^- + b - b_M^-)|^2 \nabla(u_M^- + b - b_M^-) : \nabla \partial_t b \, dx \, ds$$

$$+ \int_{d_M^-(t_1)}^{d_M(t_2)} \int_\Omega \partial_e W^{\mathrm{el}}(c_M^-, \epsilon(u_M^- + b - b_M^-), z_M^-) : e(\partial_t b) \, dx \, ds$$

$$+ \underbrace{\int_{d_M^-(t_1)}^{d_M(t_2)} \int_\Omega \partial_c W^{\mathrm{el}}(\widehat{c}_M, \epsilon(u_M^- + b_M - b_M^-), z_M^-) \partial_t \widehat{c}_M \, dx \, ds}_{(\star)}$$

$$+ \underbrace{\int_{d_M^-(t_1)}^{d_M(t_2)} \int_\Omega \left(\nabla \widehat{c}_M \cdot \nabla \partial_t \widehat{c}_M + \partial_c W^{\mathrm{ch}}(\widehat{c}_M)\partial_t \widehat{c}_M\right) dx \, ds}_{(\star\star)}$$

$$+ \underbrace{\int_{d_M^-(t_1)}^{d_M(t_2)} \int_\Omega \left(\partial_z W^{\mathrm{el}}(c_M, \epsilon(u_M^- + b_M - b_M^-), \widehat{z}_M) \partial_t \widehat{z}_M + |\nabla \widehat{z}_M|^{p-2}\nabla \widehat{z}_M \cdot \nabla \partial_t \widehat{z}_M\right) dx \, ds}_{(\star\star\star)}.$$

By using convexity of $x \mapsto |x|^p$, we obtain the following elementary inequality

$$(|\nabla \widehat{z}_M(t,x)|^{p-2}\nabla \widehat{z}_M(t,x) - |\nabla z_M(t,x)|^{p-2}\nabla z_M(t,x)) \cdot \nabla \partial_t \widehat{z}_M(t,x) \leq 0.$$

This estimate and (4.29), tested with $\zeta = -\partial_t \widehat{z}_M(t)$ for $\nu = \tau$ and integrated from $t = 0$ to $t = T$, lead to the estimate:

$$(\star\star\star) \leq -\int_{d_M^-(t_1)}^{d_M(t_2)} \int_\Omega -\alpha \partial_t \widehat{z}_M + \beta|\partial_t \widehat{z}_M|^2 \, dx \, ds$$

$$+ \underbrace{\int_{d_M^-(t_1)}^{d_M(t_2)} \int_\Omega (\partial_z W^{\mathrm{el}}(c_M, \epsilon(u_M^- + b_M - b_M^-), \widehat{z}_M) - \partial_z W^{\mathrm{el}}(c_M, \epsilon(u_M), z_M))\partial_t \widehat{z}_M \, dx \, ds}_{=:\kappa_M^3}.$$

Furthermore,

$$
\begin{aligned}
(\star) \le{} & \int_{d_M^-(t_1)}^{d_M(t_2)} \int_\Omega \partial_c W^{\mathrm{el}}(c_M, \epsilon(u_M), z_M) \partial_t \widehat{c}_M \, dx \, ds \\
& + \underbrace{\int_{d_M^-(t_1)}^{d_M(t_2)} \int_\Omega (\partial_c W^{\mathrm{el}}(\widehat{c}_M, \epsilon(u_M^- + b_M - b_M^-), z_M^-) - \partial_c W^{\mathrm{el}}(c_M, \epsilon(u_M), z_M)) \partial_t \widehat{c}_M \, dx \, ds}_{=: \kappa_M^1} \, .
\end{aligned}
$$

Using the elementary estimate $(\nabla \widehat{c}_M - \nabla c_M) \nabla \partial_t \widehat{c}_M \le 0$, we obtain

$$
\begin{aligned}
(\star\star) \le{} & \int_{d_M^-(t_1)}^{d_M(t_2)} \int_\Omega (\nabla c_M \cdot \nabla \partial_t \widehat{c}_M + \partial_c W^{\mathrm{ch}}(c_M) \partial_t \widehat{c}_M) \, dx \, ds \\
& + \underbrace{\int_{d_M^-(t_1)}^{d_M(t_2)} \int_\Omega (\partial_c W^{\mathrm{ch}}(\widehat{c}_M) - \partial_c W^{\mathrm{ch}}(c_M)) \partial_t \widehat{c}_M \, dx \, ds}_{=: \kappa_M^2} \, .
\end{aligned}
$$

Hence, applying equations (4.27) and (4.26) shows

$$
\begin{aligned}
\int_{d_M^-(t_1)}^{d_M(t_2)} \langle \mathrm{d}_c \widetilde{\mathcal{E}}_\varepsilon(q_M), \partial_t \widehat{c}_M \rangle_{(H^1)^* \times H^1} \, ds &= \int_{d_M^-(t_1)}^{d_M(t_2)} \int_\Omega \left(\mu_M \partial_t \widehat{c}_M - \varepsilon |\partial_t \widehat{c}_M|^2 \right) dx \, ds \\
&= \int_{d_M^-(t_1)}^{d_M(t_2)} \int_\Omega \left(-|\nabla \mu_M|^2 - \varepsilon |\partial_t \widehat{c}_M|^2 \right) dx \, ds.
\end{aligned}
$$

Thus,

$$
(\star) + (\star\star) \le \int_{d_M^-(t_1)}^{d_M(t_2)} \int_\Omega (-|\nabla \mu_M|^2 - \varepsilon |\partial_t \widehat{c}_M|^2) \, dx \, ds + \kappa_M^1 + \kappa_M^2 \, .
$$

In the following, we prove that the error terms κ_M^1, κ_M^2 and κ_M^3 converge to 0 as $M \to \infty$.

- By Corollary 4.3.9 and the regularity of b, we obtain

$$
\partial_c W^{\mathrm{el}}(\widehat{c}_M, \epsilon(u_M^- + b_M - b_M^-), z_M^-) - \partial_c W^{\mathrm{el}}(c_M, \epsilon(u_M), z_M) \to 0 \text{ a.e. in } \Omega_T.
$$

Growth assumption (4.2d) yields

$$
\begin{aligned}
& \left| \partial_c W^{\mathrm{el}}(\widehat{c}_M, \epsilon(u_M^- + b_M - b_M^-), z_M^-) - \partial_c W^{\mathrm{el}}(c_M, \epsilon(u_M), z_M) \right|^2 \\
& \le \left| \partial_c W^{\mathrm{el}}(\widehat{c}_M, \epsilon(u_M^- + b_M - b_M^-), z_M^-) \right| + \left| \partial_c W^{\mathrm{el}}(c_M, \epsilon(u_M), z_M) \right|^2 \\
& \le C(|\widehat{c}_M|^4 + |c_M|^4 + |\epsilon(u_M^-)|^2 + |\epsilon(u_M)|^2 + |\epsilon(b_M)|^2 + |\epsilon(b_M^-)|^2 + 1).
\end{aligned}
$$

Since the right hand side converges strongly in $L^1(\Omega_T)$ due to Corollary 4.3.9, Lebesgue's generalized convergence theorem shows

$$
\partial_c W^{\mathrm{el}}(\widehat{c}_M, \epsilon(u_M^- + b_M - b_M^-), z_M^-) - \partial_c W^{\mathrm{el}}(c_M, \epsilon(u_M), z_M) \to 0 \text{ in } L^2(\Omega_T).
$$

Using also $\partial_t \widehat{c}_M \rightharpoonup \partial_t c$ in $L^2(\Omega_T)$, we conclude $\kappa_M^1 \to 0$ as $M \to \infty$.

- Corollary 4.3.9 and the continuous embedding $H^1(\Omega) \hookrightarrow L^{2^*}(\Omega)$ yield $c_M, \widehat{c}_M \to c$ in $L^{2^*}(\Omega_T)$. Together with the pointwise a.e. convergence $c_M, \widehat{c}_M \to c$ in Ω_T and the growth condition (4.2g), i.e.,

$$\left| \partial_c W^{\mathrm{ch}}(\widehat{c}_M) - \partial_c W^{\mathrm{ch}}(c_M) \right|^2 \leq C(|\widehat{c}_M|^{2^*} + |c_M|^{2^*} + 1),$$

Lebesgue's generalized convergence theorem proves

$$\partial_c W^{\mathrm{ch}}(\widehat{c}_M) - \partial_c W^{\mathrm{ch}}(c_M) \to 0 \text{ in } L^2(\Omega_T).$$

Combining this with $\partial_t \widehat{c}_M \rightharpoonup \partial_t c$ in $L^2(\Omega_T)$, we end up with $\kappa_M^2 \to 0$ as $M \to \infty$.

- As before, by Corollary 4.3.9,

$$\partial_z W^{\mathrm{el}}(c_M, \epsilon(u_M^- + b_M - b_M^-), \widehat{z}_M) - \partial_z W^{\mathrm{el}}(c_M, \epsilon(u_M), z_M) \to 0 \text{ a.e. in } \Omega_T.$$

Growth assumption (4.2f) reveals

$$\left| \partial_z W^{\mathrm{el}}(c_M, \epsilon(u_M^- + b_M - b_M^-), \widehat{z}_M) - \partial_z W^{\mathrm{el}}(c_M, \epsilon(u_M), z_M) \right|^2$$
$$\leq C(|c_M|^4 + |\epsilon(u_M)|^4 + |\epsilon(u_M^-)|^4 + |\epsilon(b_M)|^4 + |\epsilon(b_M^-)|^4 + 1).$$

By Corollary 4.3.9 and the regularity of b, the right hand side converges strongly in $L^1(\Omega_T)$. Lebesgue's generalized convergence theorem gives

$$\partial_z W^{\mathrm{el}}(c_M, \epsilon(u_M^- + b_M - b_M^-), \widehat{z}_M) - \partial_z W^{\mathrm{el}}(c_M, \epsilon(u_M), z_M) \to 0 \text{ in } L^2(\Omega_T).$$

Using also $\partial_t \widehat{z}_M \rightharpoonup \partial_t z$ in $L^2(\Omega_T)$ by Corollary 4.3.9, we obtain $\kappa_M^3 \to 0$ as $M \to \infty$.

At this point, we want to emphasize that we have used the convergences $\epsilon(u_M), \epsilon(u_M^-) \to \epsilon(u)$ in $L^4(\Omega_T; \mathbb{R}^{n \times n})$ and $\partial_t \widehat{c}_M \rightharpoonup \partial_t c$ in $L^2(\Omega_T)$ as $M \to \infty$, which we have only due to the regularization for every fixed $\varepsilon > 0$.

To finish the proof, set $\kappa_M := \kappa_M^1 + \kappa_M^2 + \kappa_M^3$. $\qquad\square$

We are now in the position to prove the existence theorem for the viscous case.

Proof of Theorem 4.2.5. The proof is divided into several steps:

(i) Using growth conditions (4.2d), (4.2g), (4.3), Corollary 4.3.9 and Lebesgue's generalized convergence theorem, we can pass to $M \to \infty$ in the time integrated version of the integral equations (4.26), (4.27) and (4.28). This shows (i) and (ii) of Definition 4.2.3.

(ii) Let $0 \leq t_1 < t_2 \leq T$ be arbitrary. Because of $d_M^-(t_1) \leq t_1 < t_2 \leq d_M(t_2)$, Lemma 4.3.10 particularly implies

$$\mathcal{E}_\varepsilon(q_M(t_2)) + \int_{t_1}^{t_2} \int_\Omega \left(-\alpha \partial_t \widehat{z}_M + \beta |\partial_t \widehat{z}_M|^2 + \varepsilon |\partial_t \widehat{c}_M|^2 + |\nabla \mu_M|^2 \right) \mathrm{d}x \, \mathrm{d}t - \mathcal{E}_\varepsilon(q_M^-(t_1))$$

$$\leq \int_{d_M^-(t_1)}^{d_M(t_2)} \int_\Omega \partial_e W^{\mathrm{el}}(c_M^-, \epsilon(u_M^- + b - b_M^-), z_M) : \epsilon(\partial_t b) \, dx \, dt$$

$$+ \varepsilon \int_{d_M^-(t_1)}^{d_M(t_2)} \int_\Omega |\nabla u_M^- + \nabla b - \nabla b_M^-|^2 \nabla(u_M^- + b - b_M^-) : \nabla \partial_t b \, dx \, dt + \kappa_M$$

$$(4.46)$$

with $\kappa_M \to 0$ as $M \to \infty$.

We are going to show that for a.e. $t \in (0, T)$

$$\mathcal{E}_\varepsilon(q_M(t)) \to \mathcal{E}_\varepsilon(q(t)) \text{ and } \mathcal{E}_\varepsilon(q_M^-(t)) \to \mathcal{E}_\varepsilon(q(t)) \qquad (4.47)$$

as $M \to \infty$.

- Indeed, by Corollary 4.3.9, we have

$$W^{\mathrm{el}}(c_M, \epsilon(u_M), z_M) + W^{\mathrm{ch}}(c_M) \to W^{\mathrm{el}}(c, \epsilon(u), z) + W^{\mathrm{ch}}(c) \text{ a.e. in } \Omega_T.$$

Due to the growth conditions (4.2c) and (4.2g), we get

$$|W^{\mathrm{el}}(c_M, \epsilon(u_M), z_M) + W^{\mathrm{ch}}(c_M)| \leq C(|c_M|^2 + |c_M|^{2^*/2} + |\epsilon(u_M)|^2 + 1).$$

For a.e. $t \in (0, T)$, Corollary 4.3.9 yields the strong convergence

$$|c_M(t)|^2 + |c_M(t)|^{2^*/2} + |\epsilon(u_M(t))|^2 \to |c(t)|^2 + |c(t)|^{2^*/2} + |\epsilon(u(t))|^2 \text{ in } L^1(\Omega).$$

Therefore, Lebesgue's generalized convergence theorem implies the following $L^1(\Omega)$ convergence for a.e. $t \in (0, T)$:

$$W^{\mathrm{el}}(c_M(t), \epsilon(u_M(t)), z_M(t)) + W^{\mathrm{ch}}(c_M(t)) \to W^{\mathrm{el}}(c(t), \epsilon(u(t)), z(t)) + W^{\mathrm{ch}}(c(t)).$$

Taking also $\|\nabla c_M(t)\|_{L^2(\Omega)}^2 \to \|\nabla c(t)\|_{L^2(\Omega)}^2$, $\|\nabla u_M(t)\|_{L^4(\Omega)}^4 \to \|\nabla u(t)\|_{L^4(\Omega)}^4$ and $\|\nabla z_M(t)\|_{L^p(\Omega)}^p \to \|\nabla z(t)\|_{L^p(\Omega)}^p$ (by Corollary 4.3.9) into account, we obtain

$$\mathcal{E}_\varepsilon(q_M(t)) \to \mathcal{E}_\varepsilon(q(t))$$

as $M \to \infty$. The convergence of the retarded energy $\mathcal{E}_\varepsilon(q_M^-(t)) \to \mathcal{E}_\varepsilon(q(t))$ can be shown analogously.

Now, a sequentially weakly lower semi-continuity argument based on Corollary 4.3.9 shows:

$$\liminf_{M \to \infty} \int_{t_1}^{t_2} \int_\Omega \left(-\alpha \partial_t \widehat{z}_M + \beta |\partial_t \widehat{z}_M|^2 + \varepsilon |\partial_t \widehat{c}_M|^2 + |\nabla \mu_M|^2 \right) dx \, dt$$

$$\geq \int_\Omega \alpha(z(t_1) - z(t_2)) \, dx + \int_{t_1}^{t_2} \int_\Omega \left(\beta |\partial_t z|^2 + \varepsilon |\partial_t c|^2 + |\nabla \mu|^2 \right) dx \, dt. \qquad (4.48)$$

Beyond that, growth condition (4.3) and Corollary 4.3.9 show:

$$\partial_e W^{\mathrm{el}}(c_M^- \epsilon(u_M^- + b - b_M^-), z_M) \overset{*}{\rightharpoonup} \partial_e W^{\mathrm{el}}(c, \epsilon(u), z) \quad \text{in } L^\infty(0, T; L^2(\Omega)),$$
$$|\nabla u_M^- + \nabla b - \nabla b_M^-|^2 \nabla(u_M^- + b - b_M^-) \overset{*}{\rightharpoonup} |\nabla u|^2 \nabla u \quad \text{in } L^\infty(0, T; L^{4/3}(\Omega)).$$

Since $\epsilon(\partial_t b) \in L^1(0, T; L^2(\Omega))$ and $\nabla \partial_t b \in L^1(0, T; L^4(\Omega))$, we get:

$$\int_{d_M^-(t_1)}^{d_M(t_2)} \int_\Omega \partial_e W^{\mathrm{el}}(c_M^- \epsilon(u_M^- + b - b_M^-), z_M) : \epsilon(\partial_t b) \, \mathrm{d}x \, \mathrm{d}t$$
$$\rightarrow \int_{t_1}^{t_2} \int_\Omega \partial_e W^{\mathrm{el}}(c, \epsilon(u), z) : \epsilon(\partial_t b) \, \mathrm{d}x \, \mathrm{d}t,$$

$$\int_{d_M^-(t_1)}^{d_M(t_2)} \int_\Omega |\nabla u_M^- + \nabla b - \nabla b_M^-|^2 \nabla(u_M^- + b - b_M^-) : \nabla \partial_t b \, \mathrm{d}x \, \mathrm{d}t$$
$$\rightarrow \int_{t_1}^{t_2} \int_\Omega |\nabla u|^2 \nabla u : \nabla \partial_t b \, \mathrm{d}x \, \mathrm{d}t. \tag{4.49}$$

Now, using (4.47), (4.48) and (4.49) gives (iv) of Definition 4.2.3 by passing to $M \to \infty$ in (4.46) for a subsequence.

(iii) Let $\widetilde{\zeta} \in L^p(0, T; W_-^{1,p}(\Omega)) \cap L^\infty(\Omega_T)$ be a test function with $\{\widetilde{\zeta} = 0\} \supseteq \{z = 0\}$. Applying Lemma 2.3.18 with $f = z$ and $f_M = z_M$ and $\zeta = -\widetilde{\zeta}$ gives a sequence of approximations $\{\zeta_M\}_{M \in \mathbb{N}} \subseteq L^p(0, T; W_+^{1,p}(\Omega)) \cap L^\infty(\Omega_T)$ with the properties:

$$\zeta_M \to -\widetilde{\zeta} \text{ in } L^p(0, T; W^{1,p}(\Omega)) \text{ as } M \to \infty, \tag{4.50a}$$
$$0 \le \zeta_M \le -\widetilde{\zeta} \text{ a.e. in } \Omega_T \text{ for all } M \in \mathbb{N}, \tag{4.50b}$$
$$0 \le \nu_{M,t} \zeta_M(t) \le z_M(t) \text{ a.e. in } \Omega \text{ for a.e. } t \in (0, T) \text{ and all } M \in \mathbb{N}. \tag{4.50c}$$

Furthermore, we choose a subsequence of $\{\zeta_M\}$ (we omit the subscript) such that $\zeta_M \to -\widetilde{\zeta}$ a.e. in Ω_T.

Let $\widetilde{\zeta}_M$ denote the function $-\zeta_M$. Then, multiplying estimate (4.50c) with -1 implies

$$0 \ge \nu_{M,t} \widetilde{\zeta}_M(t) \ge -z_M(t).$$

Rearranging,

$$z_M(t) \ge \nu_{M,t} \widetilde{\zeta}_M(t) + z_M(t) \ge 0.$$

Due to $z_M^- \ge z_M$ in Ω_T, we get

$$0 \le \nu_{M,t} \widetilde{\zeta}_M(t) + z_M(t) \le z_M^-(t).$$

Therefore, (4.29) holds for $\zeta = \widetilde{\zeta}_M(t)$. Integration from $t = 0$ to $t = T$ yields

$$0 \le \int_{\Omega_T} \left(|\nabla z_M|^{p-2} \nabla z_M \cdot \nabla \widetilde{\zeta}_M + \left(\partial_z W^{\mathrm{el}}(c_M, \epsilon(u_M), z_M) - \alpha + \beta(\partial_t \widehat{z}_M) \right) \widetilde{\zeta}_M \right) \mathrm{d}x. \tag{4.51}$$

By growth condition (4.2f) and by (4.50b), we have

$$|\partial_z W^{\mathrm{el}}(c_M, \epsilon(u_M), z_M)\zeta_M| \le C(|c_M|^2 + |\epsilon(u_M)|^2 + 1)$$

Applying Lebesgue's generalized convergence theorem and using Corollary 4.3.9,

$$\partial_z W^{\mathrm{el}}(c_M, \epsilon(u_M), z_M)\zeta_M \to \partial_z W^{\mathrm{el}}(c, \epsilon(u), z)\zeta \text{ in } L^1(\Omega_T).$$

Using this and again Corollary 4.3.9 and convergence property (4.50a), we can pass to $M \to \infty$ in (4.51) and end up with

$$0 \le \int_{\Omega_T} \left(|\nabla z|^{p-2}\nabla z \cdot \nabla\tilde{\zeta} + \partial_z W^{\mathrm{el}}(c, \epsilon(u), z)\tilde{\zeta} - \alpha\tilde{\zeta} + \beta(\partial_t z)\tilde{\zeta}\right) \mathrm{d}x\,\mathrm{d}t. \qquad (4.52)$$

(iv) Property (4.52) implies that

$$-\int_{\Omega} \left(|\nabla z(t)|^{p-2}\nabla z(t) \cdot \nabla\zeta + \left(\partial_z W^{\mathrm{el}}(c(t), \epsilon(u(t)), z(t)) - \alpha + \beta(\partial_t z(t))\right)\zeta\right) \mathrm{d}x \le 0$$

holds for all $\zeta \in W_-^{1,p}(\Omega)$ with $\{\zeta = 0\} \supseteq \{z(t) = 0\}$ and for a.e. $t \in (0, T)$. Applying Lemma 2.3.19 with $f = |\nabla z(t)|^{p-2}\nabla z(t)$ and $g = \partial_z W^{\mathrm{el}}(c(t), \epsilon(u(t)), z(t)) - \alpha + \beta(\partial_t z(t))$ shows

$$\int_{\Omega} \left(|\nabla z(t)|^{p-2}\nabla z(t) \cdot \nabla\zeta + \left(\partial_z W^{\mathrm{el}}(c(t), \epsilon(u(t)), z(t)) - \alpha + \beta(\partial_t z(t))\right)\zeta\right) \mathrm{d}x$$

$$\ge \int_{\{z(t)=0\}} [\partial_z W^{\mathrm{el}}(c(t), \epsilon(u(t)), z(t)) - \alpha + \beta(\partial_t z(t))]^+ \zeta \,\mathrm{d}x$$

$$\ge \int_{\{z(t)=0\}} [\partial_z W^{\mathrm{el}}(c(t), \epsilon(u(t)), z(t))]^+ \zeta \,\mathrm{d}x \qquad (4.53)$$

for all $\zeta \in W_-^{1,p}(\Omega)$. Setting

$$r := -\chi_{\{z=0\}} [\partial_z W^{\mathrm{el}}(c, \epsilon(u), z)]^+,$$

we get (4.19) from (4.53) by integration from $t = 0$ to $t = T$ and we also have

$$\langle r(t), \zeta - z(t)\rangle = -\int_{\{z(t)=0\}} [\partial_z W^{\mathrm{el}}(c(t), \epsilon(u(t)), z(t))]^+ (\zeta - z(t)) \,\mathrm{d}x \le 0$$

for any $\zeta \in W_+^{1,p}(\Omega)$ and a.e. $t \in (0, T)$. Therefore, (4.20) is shown. $\qquad\square$

4.3.2   Existence of weak solutions for the limit system

For each $\varepsilon \in (0,1)$, we denote with $q_\varepsilon = (c_\varepsilon, u_\varepsilon, z_\varepsilon)$ a viscous solution according to Theorem 4.2.5. Whenever we refer to the equations and inequalities (4.16)-(4.22) of Definition 4.2.3 the variables $q = (c, u, z)$, μ and r should be replaced by $q_\varepsilon = (c_\varepsilon, u_\varepsilon, z_\varepsilon)$, μ_ε and r_ε. By the use of Lemma 4.3.11, Lemma 4.3.12 and Lemma 4.3.13 below, we identify a suitable subsequence where we can pass to the limit.

Lemma 4.3.11 (A-priori estimates) *There exists a $C > 0$ independent of $\varepsilon > 0$ such that*

(i) $\|u_\varepsilon\|_{L^\infty(0,T;H^1(\Omega;\mathbb{R}^n))} \leq C$,

(ii) $\varepsilon^{1/4}\|\nabla u_\varepsilon\|_{L^\infty(0,T;L^4(\Omega;\mathbb{R}^{n\times n}))} \leq C$,

(iii) $\|c_\varepsilon\|_{L^\infty(0,T;H^1(\Omega))} \leq C$,

(iv) $\|z_\varepsilon\|_{L^\infty(0,T;W^{1,p}(\Omega))} \leq C$,

(v) $\|\partial_t c_\varepsilon\|_{L^2(0,T;(H^1(\Omega))^*)} \leq C$,

(vi) $\varepsilon^{1/2}\|\partial_t c_\varepsilon\|_{L^2(\Omega_T)} \leq C$,

(vii) $\|\partial_t z_\varepsilon\|_{L^2(\Omega_T)} \leq C$,

(viii) $\|\mu_\varepsilon\|_{L^2(0,T;H^1(\Omega))} \leq C$

for all $\varepsilon \in (0,1)$.

Proof. According to Lemma 4.3.3, the discretization $q_{M,\varepsilon}$ of q_ε fulfills

$$\mathcal{E}_\varepsilon(q_{M,\varepsilon}(t)) + \int_0^{d_M(t)} \mathcal{R}(\partial_t \widehat{z}_{M,\varepsilon})\,ds + \int_0^{d_M(t)} \int_\Omega \left(\frac{\varepsilon}{2}|\partial_t \widehat{c}_{M,\varepsilon}|^2 + \frac{1}{2}|\nabla \mu_{M,\varepsilon}|^2\right)dx\,ds$$
$$\leq C(\mathcal{E}_\varepsilon(q_\varepsilon^0) + 1), \tag{4.54}$$

where C is independent of M, t, ε. By the minimizing property of q_ε^0, we also obtain $\mathcal{E}_\varepsilon(q_\varepsilon^0) \leq \mathcal{E}_\varepsilon(q_1^0) \leq \mathcal{E}_1(q_1^0)$ for all $\varepsilon \in (0,1)$. Therefore, the left hand side of (4.54) is bounded with respect to $M \in \mathbb{N}$, $t \in [0,T]$ and $\varepsilon \in (0,1)$. By Corollary 4.3.9, this leads to the boundedness of

$$\mathcal{E}_\varepsilon(q_\varepsilon(t)) + \int_0^t \mathcal{R}(\partial_t z_\varepsilon)\,ds + \int_0^t \int_\Omega \left(\frac{\varepsilon}{2}|\partial_t c_\varepsilon|^2 + \frac{1}{2}|\nabla \mu_\varepsilon|^2\right)dx\,ds \leq C \tag{4.55}$$

for a.e. $t \in (0,T)$ and for all $\varepsilon \in (0,1)$.

We can conclude as follows:

- We immediately obtain (ii), (iv), (vi) and (vii) from (4.55). Note that $z_\varepsilon \in [0,1]$ a.e. in Ω_T.

- Due to $\int c_\varepsilon(t)\,dx = $ const and the boundedness of $\|\nabla c_\varepsilon(t)\|_{L^2(\Omega)}$, Poincaré's inequality yields (iii).

- Now, using (4.55), growth conditions (4.4) and Korn's inequality, we attain the desired a-priori estimate (i).

- Testing (4.17) and (4.16) with $\zeta \equiv 1$, we get

$$\int_\Omega \mu_\varepsilon(t)\,dx = \int_\Omega \left(\partial_c W^{\mathrm{ch}}(c_\varepsilon(t)) + \partial_c W^{\mathrm{el}}(c_\varepsilon(t), \epsilon(u_\varepsilon(t)), z_\varepsilon(t))\right)dx$$

$$+ \underbrace{\int_\Omega \varepsilon(\partial_t c_\varepsilon(t))\,dx}_{=0 \text{ by (4.16) with } \zeta \equiv 1}$$

$$= \int_\Omega \left(\partial_c W^{\text{ch}}(c_\varepsilon(t)) + \partial_c W^{\text{el}}(c_\varepsilon(t), \epsilon(u_\varepsilon(t)), z_\varepsilon(t)) \right) dx.$$

Using the already known boundedness properties, we obtain boundedness of $\left| \int_\Omega \mu_\varepsilon(t)\,dx \right|$ w.r.t. t and ε. Since $\|\nabla \mu_\varepsilon(t)\|_{L^2(\Omega_T)}$ is also bounded, Poincaré's inequality yields (viii).

- Finally, we know from the boundedness of $\{\nabla \mu_\varepsilon\}$ in $L^2(\Omega_T)$ that $\{\partial_t c_\varepsilon\}$ is also bounded in $L^2(0,T;(H^1(\Omega))^*)$ with respect to ε by using equation (4.16). Therefore, (v) holds.

$\square$

Lemma 4.3.12 (Weak convergence of viscous solutions)

There exists a subsequence $\{\varepsilon_k\}$ (which is also denoted by ε) and functions (u, c, z, μ) satisfying Definition 4.2.6 (i) such that

(i) $z_\varepsilon \overset{*}{\rightharpoonup} z$ in $L^\infty(0,T;W^{1,p}(\Omega))$,
   $z_\varepsilon(t) \rightharpoonup z(t)$ in $W^{1,p}(\Omega)$ for a.e. t,
   $z_\varepsilon \to z$ a.e. in Ω_T
   $z_\varepsilon \rightharpoonup z$ in $H^1(0,T;L^2(\Omega))$,

(ii) $u_\varepsilon \overset{*}{\rightharpoonup} u$ in $L^\infty(0,T;H^1(\Omega;\mathbb{R}^n))$,

(iii) $c_\varepsilon \overset{*}{\rightharpoonup} c$ in $L^\infty(0,T;H^1(\Omega))$,
   $c_\varepsilon(t) \rightharpoonup c(t)$ in $H^1(\Omega)$ for a.e. t,
   $c_\varepsilon \to c$ a.e. in Ω_T,

(iv) $\mu_\varepsilon \rightharpoonup \mu$ in $L^2(0,T;H^1(\Omega))$

as $\varepsilon \to 0^+$.

Proof.

(i) This property follows from the boundedness of $\{z_\varepsilon\}$ in $L^\infty(0,T;W^{1,p}(\Omega))$ and in $H^1(0,T;L^2(\Omega))$ (see proof of Lemma 4.3.11). The function z obtained in this way is monotonically decreasing with respect to t, i.e., $\partial_t z \leq 0$ a.e. in Ω_T.

(ii) This property follows from the boundedness of $\{u_\varepsilon\}$ in $L^\infty(0,T;H^1(\Omega;\mathbb{R}^n))$.

(iii) Properties (iii) and (v) of Lemma 4.3.11 show that c_ε converges strongly to an element c in $L^2(\Omega_T)$ as $\varepsilon \to 0^+$ for a subsequence by a compactness result due to J. P. Aubin and J. L. Lions (see Theorem 2.3.9 (i)). This allows us to extract a further subsequence such that $c_\varepsilon(t) \to c(t)$ in $L^2(\Omega)$ for a.e. $t \in (0,T)$. Taking also the boundedness of $\{c_\varepsilon\}$ in $L^\infty(0, T; H^1(\Omega))$ into account, we obtain a subsequence with $c_\varepsilon(t) \rightharpoonup c(t)$ in $H^1(\Omega)$ for a.e. $t \in (0, T)$ and $c_\varepsilon \to c$ a.e. in Ω_T as well as $c_\varepsilon \overset{\star}{\rightharpoonup} c$ in $L^\infty(0, T; H^1(\Omega))$.

(iv) This property follows from the boundedness of $\{\mu_\varepsilon\}$ in $L^2(0, T; H^1(\Omega))$. $\qquad\square$

Lemma 4.3.13 (Strong convergence of viscous solutions) *The following convergence properties are satisfied for a subsequence $\varepsilon \to 0^+$:*

(i) $u_\varepsilon \to u$ in $L^2(0, T; H^1(\Omega; \mathbb{R}^n))$, $\qquad$ (iii) $z_\varepsilon \to z$ in $L^p(0, T; W^{1,p}(\Omega))$.

(ii) $c_\varepsilon \to c$ in $L^2(0, T; H^1(\Omega))$,

Proof.

(i) The following argumentation requires the density result

$$L^4(0, T; W_{\Gamma_D}^{1,4}(\Omega; \mathbb{R}^n)) \text{ is dense in } L^2(0, T; H_{\Gamma_D}^1(\Omega; \mathbb{R}^n)). \qquad (4.56)$$

To prove this property, we use the following two density results (for notational convenience, we write $\Gamma := \Gamma_D$ and $\Gamma_T := (\Gamma_D)_T$):

- It holds

$$\mathcal{D}(\overline{\Omega_T}, \Gamma_T) \text{ is dense in } H_{\Gamma_T}^1(\Omega_T) \text{ w.r.t. } H^1(\Omega_T) \text{ topology} \qquad (4.57)$$

  with

$$\mathcal{D}(\overline{\Omega_T}, \Gamma_T) := \{v \in \mathcal{C}^\infty(\overline{\Omega_T}) \,|\, v = 0 \text{ in a neighborhood of } \Gamma_T\}.$$

  This result can be found in [Ber11, Theorem 3.1] (with the Lipschitz cylinder Ω_T and Dirichlet boundary Γ_T). Here, we need that Γ has finitely many path-connected components, see Section 2.1.

- It holds

$$H^1(0, T; X) \text{ is dense in } L^2(0, T; X) \text{ w.r.t. } L^2(0, T; X) \text{ topology} \qquad (4.58)$$

  with the Banach space

$$X := H_\Gamma^1(\Omega).$$

  This standard results can be obtained by a mollifying technique (for instance, see Theorem 1.3.3 in [Vra03]).

Note that we have the isomorphism

$$H_{\Gamma_T}^1(\Omega_T) \cong L^2(0, T; H_\Gamma^1(\Omega)) \cap H^1(0, T; L^2(\Omega)),$$

where the Bochner space $L^2(0, T; H_\Gamma^1(\Omega)) \cap H^1(0, T; L^2(\Omega))$ is equipped with the norm $\| \cdot \| := \| \cdot \|_{L^2(H^1)} + \| \cdot \|_{H^1(L^2)}$.

Taking also (4.57) and (4.58) into account and using

$$H^1(0, T; H_\Gamma^1(\Omega)) \subseteq L^2(0, T; H_\Gamma^1(\Omega)) \cap H^1(0, T; L^2(\Omega)),$$

we get the following result:

$\mathcal{D}(\overline{\Omega_T}, \Gamma_T)$ is dense in $L^2(0, T; H_\Gamma^1(\Omega))$ w.r.t. $L^2(0, T; H_\Gamma^1(\Omega))$ topology.

In particular, (4.56) is proven. Using this density result, we can take an approximation sequence $\{\widetilde{u}_\delta\}_{\delta \in (0,1)} \subseteq L^4(0, T; W^{1,4}(\Omega; \mathbb{R}^n))$ with

$$\widetilde{u}_\delta \to u \text{ in } L^2(0, T; H^1(\Omega; \mathbb{R}^n)) \text{ as } \delta \to 0^+, \tag{4.59a}$$

$$\widetilde{u}_\delta - b \in L^4(0, T; W_{\Gamma_D}^{1,4}(\Omega; \mathbb{R}^n)) \text{ for all } \delta > 0. \tag{4.59b}$$

Since ε and δ are independently chosen, we find a reparametrization sequence $\{\delta_\varepsilon\}_{\varepsilon \in (0,1)}$ with $\delta_\varepsilon \to 0^+$ such that

$$\varepsilon^{1/4} \|\nabla \widetilde{u}_{\delta_\varepsilon}\|_{L^4(\Omega_T; \mathbb{R}^{n \times n})} \to 0 \text{ as } \varepsilon \to 0^+. \tag{4.60}$$

- In the following, we give a construction of the sequence $\{\delta_\varepsilon\}$. We consider the interesting case $\lim \sup_{\delta \to 0^+} \|\nabla \widetilde{u}_\delta\|_{L^4(\Omega_T; \mathbb{R}^{n \times n})} = \infty$ (otherwise, (4.60) follows with $\delta_\varepsilon := \varepsilon$). Let $\{\eta_k\}_{k \in \mathbb{N}}$ be a sequence with $\eta_k \to 0^+$ such that $a_k := \|\nabla \widetilde{u}_{\eta_k}\|_{L^4(\Omega_T; \mathbb{R}^{n \times n})} \to \infty$ as $k \to \infty$. Define the function

$$v(k) := \left(\frac{1}{a_k + 1} \right)^8.$$

Since $v(k) \to 0^+$ as $k \to \infty$, there are only finitely many values k for every $\varepsilon \in (0, 1)$ such that $v(k) \geq \varepsilon$.

Let $\varepsilon > 0$ be so small such that there exists a $k \in \mathbb{N}$ with $v(k) \geq \varepsilon$. Now, the following definition makes sense:

$$w(\varepsilon) := \max\{k \in \mathbb{N} \,|\, v(k) \geq \varepsilon\}$$

This definition implies $v(w(\varepsilon)) \geq \varepsilon$. Furthermore, the convergence $v(k) \to 0^+$ as $k \to \infty$ shows $w(\varepsilon) \to \infty$ as $\varepsilon \to 0^+$.

Using these properties and setting $\delta_\varepsilon := \eta_{w(\varepsilon)}$, we end up with the desired convergence:

$$\varepsilon^{1/4} \|\nabla \widetilde{u}_{\delta_\varepsilon}\|_{L^4(\Omega_T; \mathbb{R}^{n \times n})} \leq v(w(\varepsilon))^{1/4} \|\nabla \widetilde{u}_{\eta_{w(\varepsilon)}}\|_{L^4(\Omega_T; \mathbb{R}^{n \times n})}$$

$$= \left(\frac{1}{a_{w(\varepsilon)} + 1} \right)^{8/4} a_{w(\varepsilon)}$$

$$= \frac{a_{w(\varepsilon)}}{(a_{w(\varepsilon)} + 1)^2} \to 0^+ \text{ as } \varepsilon \to 0^+.$$

Hence, (4.60) is proven.

Testing (4.18) with $\zeta = u_\varepsilon - \widetilde{u}_{\delta_\varepsilon}$ (possible due to (4.59b)), applying the uniform monotonicity of $\partial_e W^{\mathrm{el}}$ (assumption (4.2a)) and (4.37) for $p = 4$ (compare with the calculation performed in (4.39)) gives

$$\frac{\eta}{2}\|\epsilon(u_\varepsilon) - \epsilon(u)\|_{L^2(\Omega_T)}^2$$

$$\leq \eta\|\epsilon(u) - \epsilon(\widetilde{u}_{\delta_\varepsilon})\|_{L^2(\Omega_T)}^2 + \eta\|\epsilon(u_\varepsilon) - \epsilon(\widetilde{u}_{\delta_\varepsilon})\|_{L^2(\Omega_T)}^2 + \varepsilon C_{\mathrm{uc}}\|\nabla u_\varepsilon - \nabla \widetilde{u}_{\delta_\varepsilon}\|_{L^4(\Omega_T)}^4$$

$$\leq \eta\|\epsilon(u) - \epsilon(\widetilde{u}_{\delta_\varepsilon})\|_{L^2(\Omega_T)}^2$$
$$+ \int_{\Omega_T} (\partial_e W^{\mathrm{el}}(c_\varepsilon, \epsilon(u_\varepsilon), z_\varepsilon) - \partial_e W^{\mathrm{el}}(c_\varepsilon, \epsilon(\widetilde{u}_{\delta_\varepsilon}), z_\varepsilon)) : (\epsilon(u_\varepsilon) - \epsilon(\widetilde{u}_{\delta_\varepsilon}))\,dx\,dt$$
$$+ \varepsilon \int_{\Omega_T} (|\nabla u_\varepsilon|^2 \nabla u_\varepsilon - |\nabla \widetilde{u}_{\delta_\varepsilon}|^2 \nabla \widetilde{u}_{\delta_\varepsilon}) : (\nabla u_\varepsilon - \nabla \widetilde{u}_{\delta_\varepsilon})\,dx\,dt$$

$$= \eta\|\epsilon(u) - \epsilon(\widetilde{u}_{\delta_\varepsilon})\|_{L^2(\Omega_T)}^2$$
$$+ \underbrace{\int_{\Omega_T} \left(\partial_e W^{\mathrm{el}}(c_\varepsilon, \epsilon(u_\varepsilon), z_\varepsilon) : (\epsilon(u_\varepsilon) - \epsilon(\widetilde{u}_{\delta_\varepsilon})) + \varepsilon|\nabla u_\varepsilon|^2 \nabla u_\varepsilon : (\nabla u_\varepsilon - \nabla \widetilde{u}_{\delta_\varepsilon})\right)dx\,dt}_{=0 \text{ by } (4.18)}$$
$$- \int_{\Omega_T} \partial_e W^{\mathrm{el}}(c_\varepsilon, \epsilon(\widetilde{u}_{\delta_\varepsilon}), z_\varepsilon) : (\epsilon(u_\varepsilon) - \epsilon(\widetilde{u}_{\delta_\varepsilon}))\,dx\,dt$$
$$- \underbrace{\varepsilon \int_{\Omega_T} |\nabla \widetilde{u}_{\delta_\varepsilon}|^2 \nabla \widetilde{u}_{\delta_\varepsilon} : (\nabla u_\varepsilon - \nabla \widetilde{u}_{\delta_\varepsilon})\,dx\,dt}_{(\star)} . \tag{4.61}$$

Finally,

$$|(\star)| \leq \varepsilon\|\nabla \widetilde{u}_{\delta_\varepsilon}\|_{L^4(\Omega_T)}^3 \|\nabla u_\varepsilon - \nabla \widetilde{u}_{\delta_\varepsilon}\|_{L^4(\Omega_T)}$$
$$\leq \left(\underbrace{\varepsilon^{1/4}\|\nabla \widetilde{u}_{\delta_\varepsilon}\|_{L^4(\Omega_T)}}_{\to 0 \text{ as } \varepsilon \to 0^+ \text{ by } (4.60)}\right)^3 \left(\underbrace{\varepsilon^{1/4}\|\nabla u_\varepsilon\|_{L^4(\Omega_T)}}_{\leq C \text{ by Lemma 4.3.11}} + \underbrace{\varepsilon^{1/4}\|\nabla \widetilde{u}_{\delta_\varepsilon}\|_{L^4(\Omega_T)}}_{\to 0 \text{ as } \varepsilon \to 0^+ \text{ by } (4.60)}\right).$$

From growth condition (4.3), Lemma 4.3.12 and Lebesgue's generalized convergence theorem, we obtain

$$\partial_e W^{\mathrm{el}}(c_\varepsilon, \epsilon(\widetilde{u}_{\delta_\varepsilon}), z_\varepsilon) \to \partial_e W^{\mathrm{el}}(c, \epsilon(u), z) \text{ in } L^2(\Omega_T; \mathbb{R}^{n\times n})$$

for a subsequence $\varepsilon \to 0^+$. By $u_\varepsilon \overset{*}{\rightharpoonup} u$ in $L^\infty(0, T; H^1(\Omega; \mathbb{R}^n))$ for a subsequence $\varepsilon \to 0^+$ (Lemma 4.3.12 (iii)) as well as (4.59a), we also have

$$\epsilon(u_\varepsilon) - \epsilon(\widetilde{u}_{\delta_\varepsilon}) \rightharpoonup 0 \text{ in } L^2(\Omega_T; \mathbb{R}^{n\times n})$$

as $\varepsilon \to 0^+$ for a subsequence. Therefore, every term on the right hand side of (4.61) converges to 0 as $\varepsilon \to 0^+$ for a subsequence. This shows $u_\varepsilon \to u$ in $L^2(0, T; H^1(\Omega; \mathbb{R}^n))$ as $\varepsilon \to 0^+$ for a subsequence by Korn's inequality.

(ii) Testing (4.17) with c_ε and c and passing to $\varepsilon \to 0^+$ for a subsequence eventually shows strong convergence $c_\varepsilon \to c$ in $L^2(0, T; H^1(\Omega))$ (see the argumentation in Lemma 4.3.7 and notice that $\int_{\Omega_T} \varepsilon(\partial_t c_\varepsilon) c_\varepsilon \, dx \, dt \le \varepsilon \|\partial_t c_\varepsilon\|_{L^2(\Omega_T)} \|c_\varepsilon\|_{L^2(\Omega_T)} \to 0$ as $\varepsilon \to 0^+$).

(iii) According to Lemma 2.3.18 with $f = \zeta = z$ and $f_M = z_{\varepsilon_M}$ (here we choose $\varepsilon_M = 1/M$) we find an approximation sequence $\{\zeta_{\varepsilon_k}\} \subseteq L^p(0, T; W_+^{1,p}(\Omega)) \cap L^\infty(\Omega_T)$ with $\varepsilon_k \to 0^+$ and the properties:

$$\zeta_{\varepsilon_k} \to z \text{ in } L^p(0, T; W^{1,p}(\Omega)) \text{ as } k \to \infty, \tag{4.62a}$$

$$0 \le \zeta_{\varepsilon_k} \le z_{\varepsilon_k} \text{ a.e. in } \Omega_T \text{ for all } k \in \mathbb{N}. \tag{4.62b}$$

We denote the subsequences also with $\{z_\varepsilon\}$ and $\{\zeta_\varepsilon\}$, respectively. The desired property $z_\varepsilon \to z$ in $L^p(0, T; W^{1,p}(\Omega))$ as $\varepsilon \to 0^+$ follows with the same estimates as in the proof of Lemma 4.3.8 by using the uniform convexity of $x \mapsto |x|^p$ and the integral inequality (4.19) with $\zeta := \zeta_\varepsilon - z_\varepsilon$ (note that $\langle r_\varepsilon, \zeta_\varepsilon - z_\varepsilon \rangle = 0$ holds by (4.22) and (4.62b)). Indeed, we obtain

$$C_{\text{ineq}}^{-1} \int_{\Omega_T} |\nabla z_\varepsilon - \nabla z|^p \, dx \, dt$$

$$\le \underbrace{\|\partial_z W^{\text{el}}(c_\varepsilon, \epsilon(u_\varepsilon), z_\varepsilon) - \alpha + \beta \partial_t z_\varepsilon\|_{L^2(0,T;L^1(\Omega))}}_{\text{bounded}} \underbrace{\|\zeta_\varepsilon - z_\varepsilon\|_{L^2(0,T;L^\infty(\Omega))}}_{\to 0}$$

$$+ \underbrace{\|\nabla z_\varepsilon\|_{L^p(\Omega_T)}^{p-1}}_{\text{bounded}} \underbrace{\|\nabla \zeta_\varepsilon - \nabla z\|_{L^p(\Omega_T)}}_{\to 0} - \underbrace{\int_{\Omega_T} |\nabla z|^{p-2} \nabla z \cdot \nabla(z_\varepsilon - z) \, dx \, dt}_{\to 0}$$

as $\varepsilon \to 0^+$ for a subsequence. Here, we have used $z_\varepsilon \to z$ and $\zeta_\varepsilon \to z$ in $L^2(0, T; L^\infty(\Omega))$ as $\varepsilon \to 0^+$ for a subsequence due to Lemma 4.3.12 and the compact embedding $W^{1,p}(\Omega) \hookrightarrow L^\infty(\Omega)$. $\qquad\square$

Corollary 4.3.14 *The following convergence properties are fulfilled:*

(i) $z_\varepsilon \to z$ in $L^p(0, T; W^{1,p}(\Omega))$,
   $z_\varepsilon(t) \to z(t)$ in $W^{1,p}(\Omega)$ for a.e. t,
   $z_\varepsilon \to z$ a.e. in Ω_T,
   $z_\varepsilon \rightharpoonup z$ in $H^1(0, T; L^2(\Omega))$,

(ii) $c_\varepsilon \to c$ in $L^{2^*}(0, T; H^1(\Omega))$,
   $c_\varepsilon(t) \to c(t)$ in $H^1(\Omega)$ for a.e. t,
   $c_\varepsilon \to c$ a.e. in Ω_T,

(iii) $u_\varepsilon \to u$ in $L^2(0, T; H^1(\Omega; \mathbb{R}^n))$,
   $u_\varepsilon(t) \to u(t)$ in $H^1(\Omega; \mathbb{R}^n)$ for a.e. t,
   $u_\varepsilon \to u$ a.e. in Ω_T,

(iv) $\mu_\varepsilon \rightharpoonup \mu$ *in* $L^2(0, T; H^1(\Omega))$,

(v) $\partial_c W^{\text{ch}}(c_\varepsilon) \to \partial_c W^{\text{ch}}(c)$ *in* $L^2(\Omega_T)$

as $\varepsilon \to 0^+$ *for a subsequence.*

Now, we are well prepared to prove the main result in this chapter.

Proof of Theorem 4.2.7. We can pass to $\varepsilon \to 0^+$ in (4.17) and (4.18) by the already known convergence features (see Corollary 4.3.14) noticing that $\int_{\Omega_T} \varepsilon |\nabla u_\varepsilon|^2 \nabla u_\varepsilon$: $\nabla \zeta \, \mathrm{d}x \, \mathrm{d}t$ and $\int_{\Omega_T} \varepsilon (\partial_t c_\varepsilon) \zeta \, \mathrm{d}x \, \mathrm{d}t$ converge to 0 as $\varepsilon \to 0^+$. We get

$$\int_{\Omega_T} \partial_e W^{\text{el}}(c, \epsilon(u), z) : \epsilon(\zeta) \, \mathrm{d}x \, \mathrm{d}t = 0 \tag{4.63}$$

for all $\zeta \in L^4(0, T; W_{\Gamma_{\mathrm{D}}}^{1,4}(\Omega; \mathbb{R}^n))$. A density argument shows that (4.63) also holds for all $\zeta \in L^2(0, T; H_{\Gamma_{\mathrm{D}}}^1(\Omega; \mathbb{R}^n))$. Writing (4.16) in the form

$$\int_{\Omega_T} (c_\varepsilon - c^0) \partial_t \zeta \, \mathrm{d}x \, \mathrm{d}t = \int_{\Omega_T} \nabla \mu_\varepsilon \cdot \nabla \zeta \, \mathrm{d}x \, \mathrm{d}t,$$

by only allowing test functions $\zeta \in L^2(0, T; H^1(\Omega))$ with $\partial_t \zeta \in L^2(\Omega_T)$ and $\zeta(T) = 0$, we can also pass to $\varepsilon \to 0^+$ by using Corollary 4.3.14.

To obtain a limit equation in (4.19) and (4.20), observe that

$$[\partial_z W^{\text{el}}(c_\varepsilon, \epsilon(u_\varepsilon), z_\varepsilon)]^+ \to [\partial_z W^{\text{el}}(c, \epsilon(u), z)]^+ \qquad \text{in } L^1(\Omega_T),$$
$$\chi_{\{z_\varepsilon = 0\}} \overset{\star}{\rightharpoonup} \chi, \qquad \text{in } L^\infty(\Omega_T)$$

for a subsequence $\varepsilon \to 0^+$ and an element $\chi \in L^\infty(\Omega_T)$. Setting $r := -\chi [\partial_z W^{\text{el}}(c, \epsilon(u), z)]^+$ and keeping (4.22) into account, we find for all $\zeta \in L^\infty(\Omega_T)$:

$$\int_{\Omega_T} r_\varepsilon \zeta \, \mathrm{d}x \, \mathrm{d}t \to \int_{\Omega_T} r \zeta \, \mathrm{d}x \, \mathrm{d}t \tag{4.64}$$

for a subsequence $\varepsilon \to 0^+$. Thus, we can also pass to $\varepsilon \to 0^+$ for a subsequence in (4.19) by using Lebesgue's generalized convergence theorem, growth condition (4.2f), Corollary 4.3.14 and (4.64). Let $\xi \in L^\infty([0, T])$ with $\xi \geq 0$ a.e. on $[0, T]$ be a further test function. Then, (4.20) and (4.22) imply

$$0 \geq \int_0^T \left(\int_\Omega r_\varepsilon(t)(\zeta - z_\varepsilon(t)) \, \mathrm{d}x \right) \xi(t) \, \mathrm{d}t = \int_{\Omega_T} r_\varepsilon(\zeta - z_\varepsilon)\xi \, \mathrm{d}x \, \mathrm{d}t$$
$$\to \int_{\Omega_T} r(\zeta - z)\xi \, \mathrm{d}x \, \mathrm{d}t = \int_0^T \left(\int_\Omega r(t)(\zeta - z(t)) \, \mathrm{d}x \right) \xi(t) \, \mathrm{d}t.$$

This shows $\int_\Omega r(t)(\zeta - z(t)) \, \mathrm{d}x \leq 0$ for a.e. $t \in (0, T)$.

It remains to show that (4.21) also yields to a limit inequality. First observe that (4.21) implies:

$$\mathcal{E}_\varepsilon(q_\varepsilon(t_2)) + \int_\Omega \alpha(z_\varepsilon(t_1) - z_\varepsilon(t_2))\,\mathrm{d}x + \int_{t_1}^{t_2} \int_\Omega \left(\beta|\partial_t z_\varepsilon|^2 + |\nabla\mu_\varepsilon|^2\right)\mathrm{d}x\,\mathrm{d}t - \mathcal{E}_\varepsilon(q_\varepsilon(t_1))$$

$$\leq \int_{t_1}^{t_2} \int_\Omega \partial_e W^{\mathrm{el}}(c_\varepsilon, \epsilon(u_\varepsilon), z_\varepsilon) : e(\partial_t b)\,\mathrm{d}x\,\mathrm{d}t + \varepsilon \int_{t_1}^{t_2} \int_\Omega |\nabla u_\varepsilon|^2 \nabla u_\varepsilon : \nabla \partial_t b\,\mathrm{d}x\,\mathrm{d}t.$$

$$(4.65)$$

To proceed, we need to prove $\varepsilon \int_\Omega |\nabla u_\varepsilon(t)|^4\,\mathrm{d}x \to 0$ as $\varepsilon \to 0^+$ for a.e. $t \in (0,T)$. Indeed, testing (4.18) with $\zeta := u_\varepsilon - b$ gives

$$\varepsilon \int_{\Omega_T} |\nabla u_\varepsilon|^4\,\mathrm{d}x\,\mathrm{d}t = \varepsilon \int_{\Omega_T} |\nabla u_\varepsilon|^2 \nabla u_\varepsilon : \nabla b\,\mathrm{d}x\,\mathrm{d}t - \int_{\Omega_T} \partial_e W^{\mathrm{el}}(c_\varepsilon, \epsilon(u_\varepsilon), z_\varepsilon) : \epsilon(u_\varepsilon - b)\,\mathrm{d}x\,\mathrm{d}t.$$

We immediately see that the first term converges to 0 as $\varepsilon \to 0^+$. The second term also converges to 0 because of $\int_{\Omega_T} \partial_e W^{\mathrm{el}}(c, \epsilon(u), z) : \epsilon(u - b)\,\mathrm{d}x\,\mathrm{d}t = 0$ (see equation (4.63)). This, together with Corollary 4.3.14, proves $\mathcal{E}_\varepsilon(q_\varepsilon(t)) \to \mathcal{E}(q(t))$ for a.e. $t \in (0,T)$. In conclusion, we can pass to $\varepsilon \to 0^+$ in (4.65) for a.e. $0 \leq t_1 < t_2 \leq T$ by Corollary 4.3.14 together with Lebesgue's generalized convergence theorem, growth condition (4.2c), (4.3) and (4.2g) as well as by a sequentially weakly lower semi-continuity argument for $\int_\Omega \beta|\partial_t z_\varepsilon|^2\,\mathrm{d}x$ and for $\int_\Omega |\nabla\mu_\varepsilon|^2\,\mathrm{d}x$. □

Cahn-Hilliard systems with logarithmic chemical potentials coupled with damage processes and inhomogeneous elasticity

The existence results for weak solutions from Chapter 4 are generalized to a broader class of coupled PDE systems in this chapter. More specifically, we will be able to treat

- multi-component Cahn-Hilliard systems,
- inhomogeneous elastic energy densities,
- chemical potentials of polynomial or logarithmic type,
- quadratic gradient term of the damage variable in the energy, i.e., $p = 2$ in (1.3).

Additionally, we show that the results also apply to elastic Allen-Cahn systems coupled with damage processes. This case is even easier to treat.

The results and proofs in this chapter are published in [HK13].

5.1   Assumptions

The general setting, the growth assumptions and the assumptions on the coefficient tensors which are mandatory for the existence theorems are summarized below (see Section 3.4 for an explanation).

(i) *Energy density functions and gradient tensors.*

$$W^{\text{el}} \in \mathcal{C}^1(\mathbb{R}^N \times \mathbb{R}^{n \times n} \times \mathbb{R}; \mathbb{R}_+), \tag{5.1a}$$

$$W^{\text{ch,pol}} \in \mathcal{C}^1(\mathbb{R}^N; \mathbb{R}_+), \quad W^{\text{ch,log}} \in \mathcal{C}^1((0,1)^N; \mathbb{R}_+), \tag{5.1b}$$

$$\mathbb{M} \in \mathbb{R}^{N \times N} \text{ symmetric and positive definite on } \{x \in \mathbb{R}^N \mid \sum_{i=1}^N x_i = 0\}$$

$$\text{and } \sum_{l=1}^N \mathbb{M}_{kl} = 0 \text{ for all } k = 1, \ldots, N, \tag{5.1c}$$

$$\Gamma \in \mathcal{L}(\mathbb{R}^{N \times n}; \mathbb{R}^{N \times n}) \text{ symmetric and positive definite}, \tag{5.1d}$$

$$f = 0, \tag{5.1e}$$

$$p = 2. \tag{5.1f}$$

(ii) *Convexity, structural and growth assumptions.*

The functions W^{el} and $W^{\text{ch,pol}}$ are assumed to satisfy for some constants $\eta > 0$ and $C > 0$ the following estimates:

$$\eta|e_1 - e_2|^2 \leq (\partial_e W^{\text{el}}(c, e_1, z) - \partial_e W^{\text{el}}(c, e_2, z)) : (e_1 - e_2), \tag{5.2a}$$

$$W^{\text{el}}(c, e, z) = W^{\text{el}}(c, e^t, z), \tag{5.2b}$$

$$W^{\text{el}}(c, e, z) \leq C(|c|^2 + |e|^2 + 1), \tag{5.2c}$$

$$|\partial_c W^{\text{el}}(c, e, z)| \leq C(|c|^2 + |e|^2 + 1), \tag{5.2d}$$

$$|\partial_e W^{\text{el}}(e_1 + e_2, c, z)| \leq C(W^{\text{el}}(c, e_1, z) + |e_2| + 1), \tag{5.2e}$$

$$|\partial_z W^{\text{el}}(c, e, z)| \leq C(|c|^2 + |e|^2 + 1), \tag{5.2f}$$

$$|\partial_c W^{\text{ch,pol}}(c)| \leq C(|c|^{2^\star/2} + 1) \tag{5.2g}$$

for arbitrary $c \in \mathbb{R}$, $z \in \mathbb{R}$ and symmetric $e, e_1, e_2 \in \mathbb{R}^{n \times n}$.

The logarithmic chemical energy density functions $W^{\text{ch,log}}$ is given by

$$W^{\text{ch,log}}(c) = \theta \sum_{k=1}^N c_k \log(c_k) + \frac{1}{2} c \cdot Ac, \quad \theta > 0, \ A \in \mathbb{R}^{n \times n}_{\text{sym}}. \tag{5.3}$$

Remark 5.1.1   *(i) As in Chapter 4, the potential f can be incorporated into the energy density W^{el}. Without loss of generality, we assume $f = 0$.*

(ii) Note that the inhomogeneous elastic energy density (1.5) fits into our setting with the growth assumptions (5.2a)-(5.2f). In particular, we are not confined to homogeneous elasticity as in Chapter 4. There, the more restrictive growth condition $|\partial_c W^{\text{el}}(c, e, z)| \leq C(|e| + |c|^2 + 1)$ is used instead of (5.2d).

This chapter also considers Allen-Cahn equations instead of Cahn-Hilliard equations in the system (3.27). In this case, the equations (3.27a)-(3.27b) are substituted by the following PDE:

$$\partial_t c = \mathbb{M}w \quad \text{with} \quad w = \mathbb{P}(-\text{div}(\partial_{\nabla c}\psi) + \partial_c\psi).$$

To unify notation, we define the operator $\mathcal{S}$ from case to case by

Allen-Cahn (AC): $\quad \mathcal{S} : L^2(\Omega; \mathbb{R}^N) \to L^2(\Omega; \mathbb{R}^N), \qquad \mathcal{S}(f) := \mathbb{M}f,$ (5.4a)

Cahn-Hilliard (CH): $\quad \mathcal{S} : H^1(\Omega; \mathbb{R}^N) \to (H^1(\Omega; \mathbb{R}^N))^*, \quad \mathcal{S}(f) := \langle\mathbb{M}\nabla f, \nabla\cdot\rangle_{L^2}.$ (5.4b)

5.2 Weak formulations and existence results

In Chapter 4, we have introduced a weak formulation of (3.27) which uses a variational inequality and an energy inequality to characterize the differential inclusion (3.27d). The weak notion, which we will derive in this section for (3.27), is slightly weaker in comparison to the notion in the previous chapter and is used to handle inhomogeneous elastic energy densities.

In the following, the corresponding free energy

$$\mathcal{E} : H^1(\Omega; \mathbb{R}^N) \times H^1(\Omega; \mathbb{R}^n) \times (H^1(\Omega) \cap L^\infty(\Omega)) \to \mathbb{R}_\infty$$

and the dissipation functional

$$\mathcal{R} : L^2(\Omega) \to \mathbb{R}_\infty$$

to system (3.27) are given by

$$\mathcal{E}(c, u, z) := \int_\Omega \psi(c, \nabla c, \epsilon(u), z, \nabla z) \, dx$$

$$= \int_\Omega \left(\frac{1}{2}\Gamma\nabla c : \nabla c + \frac{1}{2}|\nabla z|^2 + W^{\text{ch}}(c) + W^{\text{el}}(c, \epsilon(u), z) + I_{[0,\infty)}(z)\right) dx, \quad (5.5a)$$

$$\mathcal{R}(z_t) := \int_\Omega \phi(z_t) \, dx = \int_\Omega \left(-\alpha z_t + \frac{\beta}{2}|z_t|^2 + I_{(-\infty,0]}(z_t)\right) dx, \quad (5.5b)$$

with viscosity constants $\alpha, \beta > 0$. The Fréchet-differentiable functionals $\widetilde{\mathcal{E}}$ and $\widetilde{\mathcal{R}}$ are given by $\mathcal{E}$ and $\mathcal{R}$ without the indicator functions. If we equip the space $H^1(\Omega) \cap L^\infty(\Omega)$ with the norm $\|\cdot\|_{H^1 \cap L^\infty} := \|\cdot\|_{H^1} + \|\cdot\|_{L^\infty}$ the generalized subdifferential $\partial_z\mathcal{E}$ at a point $(c, u, z) \in H^1(\Omega; \mathbb{R}^N) \times H^1(\Omega; \mathbb{R}^n) \times (H^1(\Omega) \cap L^\infty(\Omega))$ is

$$\partial_z\mathcal{E}(c, u, z) = \left\{d_z\widetilde{\mathcal{E}}(c, u, z) + r \in (H^1(\Omega) \cap L^\infty(\Omega))^* \,\Big|\, r \in \partial I_{H^1_+(\Omega)\cap L^\infty(\Omega)}(z)\right\}. \quad (5.6)$$

The inclusion $L^1(\Omega) \subset (H^1(\Omega) \cap L^\infty(\Omega))^*$ will be later used for the construction of a specific subgradient. Using property (5.6), the differential inclusion in (3.27d) can be rewritten in a weaker form as

$$0 \in \partial_z \mathcal{E}(c, u, z) + \partial_{\dot{z}} \mathcal{R}(\dot{z}).$$

The analytical basis for the formulation of a weak solution in this chapter is the following proposition (cf. Proposition 4.2.1):

Proposition 5.2.1 (Energetic characterization)
Let $(c, u, z, w) \in C^2(\overline{\Omega_T}; \mathbb{R}^n \times \mathbb{R}^N \times \mathbb{R} \times \mathbb{R}^N)$ be a smooth solution satisfying (3.27a)-(3.27c) with the initial-boundary conditions (3.28). Then, the following two conditions are equivalent:

(i) $0 \in \partial_z \mathcal{E}(c(t), u(t), z(t)) + \partial_{\dot{z}} \mathcal{R}(\dot{z}(t))$ for all $t \in [0, T]$,

(ii) the energy inequality

$$\mathcal{E}(c(t), u(t), z(t)) + \int_0^t \langle d_{\dot{z}}\widetilde{\mathcal{R}}(\partial_t z), \partial_t z \rangle \, ds + \int_0^t \langle \mathcal{S}w(s), w(s) \rangle ds$$

$$\leq \mathcal{E}(u(0), c(0), z(0)) + \int_{\Omega_t} \partial_e W^{\mathrm{el}}(c, \epsilon(u), z) : e(\partial_t b) \, dx \, ds$$

for all $0 \leq t \leq T$ and the variational inequality

$$0 \leq \left\langle d_z \widetilde{\mathcal{E}}(c(t), u(t), z(t)) + r(t) + d_{\dot{z}}\widetilde{\mathcal{R}}(\partial_t z(t)), \zeta \right\rangle$$

for all $\zeta \in H^1_-(\Omega) \cap L^\infty(\Omega)$ and $r(t) \in \partial I_{H^1_+(\Omega) \cap L^\infty(\Omega)}(z(t))$ and for all $0 \leq t \leq T$.

If one of the two conditions holds then the following energy balance equation is satisfied:

$$\mathcal{E}(c(t), u(t), z(t)) + \int_0^t \langle d_{\dot{z}}\widetilde{\mathcal{R}}(\partial_t z), \partial_t z \rangle \, ds + \int_0^t \langle \mathcal{S}w(s), w(s) \rangle ds$$

$$= \mathcal{E}(u(0), c(0), z(0)) + \int_{\Omega_t} \partial_e W^{\mathrm{el}}(c, \epsilon(u), z) : e(\partial_t b) \, dx \, ds.$$

Remarks for Proposition 5.2.1. In contrast to (4.12) in Proposition 4.2.1, the energy inequality in (ii) compares the energy at the beginning $s = 0$ with the energy at an arbitrary time $s = t$ instead of $s = t_1$ with $s = t_2$ for $0 \leq t_1 < t_2 \leq T$. Applying the chain rule on the right hand side of

$$\mathcal{E}(c(t), u(t), z(t)) - \mathcal{E}(c(0), u(0), z(0)) = \int_0^t \frac{d}{dt}\widetilde{\mathcal{E}}(c(s), u(s), z(s)) \, ds$$

and using (3.27a)-(3.27c) as well as the variational inequality in (ii), the "$\geq$"-part of the energy balance can be shown.

We will see that in our approach the mathematical analysis of (3.27) with the assumptions in Section 5.1 requires several ε-regularization terms to establish the energy and variational inequality for the differential inclusion and to handle the logarithmic free energy. More precisely, we consider the regularized system (cf. (4.6))

$$\partial_t c = \operatorname{div}(\mathbf{M}\nabla\mu), \tag{5.7a}$$

$$\mu = \mathbb{P}\big(-\operatorname{div}(\Gamma\nabla c) + \partial_c W^{\mathrm{ch}}(c) + \partial_c W^{\mathrm{el}}(c, \epsilon(u), z) \big) + \varepsilon\partial_t c, \tag{5.7b}$$

$$\operatorname{div}(\partial_\epsilon W^{\mathrm{el}}(c, \epsilon(u), z)) + \varepsilon\operatorname{div}(|\nabla u|^2\nabla u) = 0, \tag{5.7c}$$

$$0 = -\varepsilon\Delta_q z - \Delta z + \partial_z W^{\mathrm{el}}(c, \epsilon(u), z) + \zeta - \alpha + \beta\partial_t z + \varrho \tag{5.7d}$$

with $\zeta \in \partial I_{[0,\infty)}(z)$ and $\varrho \in I_{(-\infty,0]}(\partial_t z)$. Here, q denotes a fixed constant with $q > n$. Notice that $\mathbb{P}\partial_t c = \partial_t c$ because of $\sum_{i=1}^N \partial_t c_i(t, x) = 0$. A transition to $\varepsilon \to 0^+$ will finally give us a solution of the limit problem (3.27).

Proposition 5.2.1 can also be formulated for the regularized system (5.7) with the regularized energy

$$\widetilde{\mathcal{E}}_\varepsilon(c, u, z) := \int_\Omega \left(\frac{1}{2}\Gamma\nabla c : \nabla c + \frac{1}{2}|\nabla z|^2 + W^{\mathrm{ch,pol}}(c) + W^{\mathrm{el}}(c, \epsilon(u), z) \right) dx$$

$$+ \varepsilon \int_\Omega \left(\frac{1}{4}|\nabla u|^4 + \frac{1}{q}|\nabla z|^q \right) dx,$$

$$\mathcal{E}_\varepsilon(c, u, z) := \widetilde{\mathcal{E}}_\varepsilon(c, u, z) + \int_\Omega I_{[0,\infty)}(z)\, dx,$$

and the initial-boundary conditions (3.28).

This motivates the following weak notion of (5.7) with (3.28).

Definition 5.2.2 (Weak solution for the regularized system (5.7),(3.28))
We call a quadruple (c, u, z, w) a weak solution of the regularized system (5.7) with the initial-boundary conditions (3.28) if the following properties are satisfied:

(i) the functions are in the following spaces:

$$c \in L^\infty(0, T; H^1(\Omega; \mathbb{R}^N)) \cap H^1(0, T; L^2(\Omega; \mathbb{R}^N)), \quad c(0) = c^0, \; c_1 + \ldots + c_N = 1,$$

$$u \in L^\infty(0, T; W^{1,4}(\Omega; \mathbb{R}^n)), \quad u|_{(\Gamma_D)_T} = b|_{(\Gamma_D)_T},$$

$$z \in L^\infty(0, T; W^{1,q}(\Omega)) \cap H^1(0, T; L^2(\Omega)), \quad z(0) = z^0, \; z \geq 0, \; \partial_t z \leq 0,$$

and

$$w \in L^2(0, T; H^1(\Omega; \mathbb{R}^N)) \qquad \text{for C-H systems,}$$

$$w \in L^2(\Omega_T; \mathbb{R}^N) \qquad\qquad \text{for A-C systems}$$

(ii) for all $\zeta \in H^1(\Omega; \mathbb{R}^N)$ and for a.e. $t \in (0, T)$:

$$\int_\Omega \partial_t c(t) \cdot \zeta \, dx = \begin{cases} \int_\Omega \mathbf{M}\nabla w(t) : \nabla\zeta \, dx & \text{for C-H systems,} \\ \int_\Omega \mathbf{M} w(t) \cdot \zeta \, dx & \text{for A-H systems} \end{cases} \tag{5.8}$$

(iii) for all $\zeta \in H^1(\Omega; \mathbb{R}^N)$ and for a.e. $t \in (0, T)$:

$$\int_\Omega w(t) \cdot \zeta \, dx = \int_\Omega \mathbb{P}\Gamma\nabla c(t) : \nabla\zeta \, dx$$
$$+ \int_\Omega \left(\mathbb{P}\partial_c W^{\mathrm{ch,pol}}(c(t)) + \mathbb{P}\partial_c W^{\mathrm{el}}(c(t), \epsilon(u(t)), z(t)) + \varepsilon\partial_t c(t) \right) \cdot \zeta \, dx$$

$$(5.9)$$

(iv) for all $\zeta \in W^{1,4}_{\Gamma_\mathrm{D}}(\Omega; \mathbb{R}^n)$ and for a.e. $t \in (0, T)$:

$$\int_\Omega \left(\partial_e W^{\mathrm{el}}(c(t), \epsilon(u(t)), z(t)) : \epsilon(\zeta) + \varepsilon|\nabla u(t)|^2 \nabla u(t) : \nabla\zeta \right) dx = 0 \qquad (5.10)$$

(v) for all $\zeta \in W^{1,q}_-(\Omega)$ and for a.e. $t \in (0, T)$:

$$0 \leq \int_\Omega (\varepsilon|\nabla z(t)|^{q-2} + 1)\nabla z(t) \cdot \nabla\zeta \, dx$$
$$+ \int_\Omega \left(\partial_z W^{\mathrm{el}}(c(t), \epsilon(u(t)), z(t)) - \alpha + \beta\partial_t z(t) + r(t) \right) \zeta \, dx, \qquad (5.11)$$

where $r \in L^1(\Omega_T) \subset L^1\left(0, T; (W^{1,q}(\Omega))^*\right)$ satisfies for all $\zeta \in W^{1,q}_+(\Omega)$ and for a.e. $t \in (0, T)$:

$$\int_\Omega r(t)(\zeta - z(t)) \leq 0, \qquad (5.12)$$

(vi) energy inequality for a.e. $t \in (0, T)$:

$$\mathcal{E}_\varepsilon(c(t), u(t), z(t)) + \int_{\Omega_t} \left(-\alpha\partial_t z + \beta|\partial_t z|^2 + \varepsilon|\partial_t c|^2 \right) dx \, ds + \int_0^t \langle \mathcal{S}w(s), w(s) \rangle \, ds$$
$$\leq \mathcal{E}_\varepsilon(c^0, u^0, z^0) + \int_{\Omega_t} \partial_e W^{\mathrm{el}}(c, \epsilon(u), z) : e(\partial_t b) \, dx \, ds + \varepsilon\int_{\Omega_t} |\nabla u|^2 \nabla u : \nabla\partial_t b \, dx \, ds,$$

$$(5.13)$$

where u^0 is the unique minimizer of $\mathcal{E}_\varepsilon(c^0, \cdot, z^0)$ in $W^{1,4}(\Omega; \mathbb{R}^n)$ with trace $u^0|_{\Gamma_\mathrm{D}} = b(0)|_{\Gamma_\mathrm{D}}$.

With the help of the operator $\mathcal{S}$ (see (5.4)), the diffusion equation (5.8) can also be written as

$$\int_\Omega \partial_t c(t) \cdot \zeta \, dx = -\langle \mathcal{S}w(t), \zeta \rangle,$$

which will be used in the following.

Definition 5.2.3 (Weak solution for the limit system (3.27)-(3.28))
A quadruple (c, u, z, w) is called a weak solution of the system (3.27) with the initial-boundary conditions (3.28) if the following properties are satisfied:

(i) the functions are in the following spaces:

$$c \in L^\infty(0, T; H^1(\Omega; \mathbb{R}^N)), \ c_1 + \ldots + c_N = 1,$$

$$u \in L^\infty(0, T; H^1(\Omega; \mathbb{R}^n)), \ u|_{(\Gamma_{\Gamma_D})_T} = b|_{(\Gamma_{\Gamma_D})_T},$$

$$z \in L^\infty(0, T; H^1(\Omega)) \cap H^1(0, T; L^2(\Omega)), \ z(0) = z^0, \ z \geq 0, \ \partial_t z \leq 0$$

and

$$w \in L^2(0, T; H^1(\Omega; \mathbb{R}^N)) \qquad \text{for C-H systems,}$$

$$w \in L^2(\Omega_T; \mathbb{R}^N) \qquad \text{for A-C systems}$$

(ii) for all $\zeta \in L^2(0, T; H^1(\Omega; \mathbb{R}^N))$ with $\partial_t \zeta \in L^2(\Omega_T; \mathbb{R}^N)$ and $\zeta(T) = 0$:

$$\int_{\Omega_T} (c - c^0) \cdot \partial_t \zeta \, dx \, dt = \int_0^T \langle \mathcal{S}w, \zeta \rangle \, dt \tag{5.14}$$

(iii) for all $\zeta \in H^1(\Omega; \mathbb{R}^N) \cap L^\infty(\Omega; \mathbb{R}^N)$ and for a.e. $t \in (0, T)$:

$$\int_\Omega w(t) \cdot \zeta \, dx = \int_\Omega \mathbb{P}\Gamma\nabla c(t) : \nabla \zeta \, dx$$
$$+ \int_\Omega \left(\mathbb{P}\partial_c W^{\text{ch}}(c(t)) + \mathbb{P}\partial_c W^{\text{el}}(c(t), \epsilon(u(t)), z(t)) \right) \cdot \zeta \, dx \tag{5.15}$$

(iv) for all $\zeta \in H^1_{\Gamma_D}(\Omega; \mathbb{R}^n)$ and for a.e. $t \in (0, T)$:

$$\int_\Omega \partial_e W^{\text{el}}(c(t), \epsilon(u(t)), z(t)) : \epsilon(\zeta) \, dx = 0 \tag{5.16}$$

(v) for all $\zeta \in H^1_-(\Omega) \cap L^\infty(\Omega)$ and for a.e. $t \in (0, T)$:

$$0 \leq \int_\Omega \left(\nabla z(t) \cdot \nabla \zeta + (\partial_z W^{\text{el}}(c(t), \epsilon(u(t)), z(t)) - \alpha + \beta \partial_t z(t) + r(t))\zeta \right) dx, \tag{5.17}$$

where $r \in L^1(\Omega_T) \subset L^1(0, T; (H^1(\Omega) \cap L^\infty(\Omega))^*)$ satisfies for all $\zeta \in H^1_+(\Omega) \cap L^\infty(\Omega)$ and for a.e. $t \in (0, T)$:

$$\int_\Omega r(t)(\zeta - z(t)) \leq 0, \tag{5.18}$$

(vi) energy inequality for a.e. $t \in (0, T)$:

$$\mathcal{E}(c(t), u(t), z(t)) + \int_{\Omega_t} \left(-\alpha \partial_t z + \beta |\partial_t z|^2 \right) dx \, ds + \int_0^t \langle \mathcal{S}w(s), w(s) \rangle \, ds$$
$$\leq \mathcal{E}(c^0, u^0, z^0) + \int_{\Omega_t} \partial_e W^{\text{el}}(c, \epsilon(u), z) : e(\partial_t b) \, dx \, ds, \tag{5.19}$$

where u^0 is the unique minimizer of $\mathcal{E}(c^0, \cdot, z^0)$ in $H^1(\Omega; \mathbb{R}^n)$ with trace $u^0|_{\Gamma_D} = b(0)|_{\Gamma_D}$.

Remark 5.2.4 *Note that both notions of weak solution imply chemical mass conservation, i.e., $\int_\Omega c(t)\,\mathrm{d}x \equiv const$ for a.e. $t \in (0, T)$.*

The main results of this chapter are summarized in the following theorems:

Theorem 5.2.5 (Existence theorem - viscous, polynomial case)
Let $q > n$ and let the assumptions (5.1) and (5.2) be satisfied. Furthermore, let $b \in W^{1,1}(0, T; W^{1,\infty}(\Omega; \mathbb{R}^n))$, $c^0 \in H^1(\Omega; \mathbb{R}^N)$ with $\sum_{i=1}^{N} c_i^0 = 1$, $z^0 \in W^{1,q}(\Omega)$ with $0 \leq z^0 \leq 1$ and a viscosity factor $\varepsilon \in (0, 1)$ be given.
Then, there exists a weak solution (c, u, z, w) in the sense of Definition 5.2.2. Additionally, $r = -\chi_{\{z=0\}}[\partial_z W^{el}(c, \epsilon(u), z)]^+$.

Theorem 5.2.6 (Existence theorem - limit, polynomial case)
Let the assumptions (5.1) and (5.2) be satisfied and let $b \in W^{1,1}(0, T; W^{1,\infty}(\Omega; \mathbb{R}^n))$, $c^0 \in H^1(\Omega; \mathbb{R}^N)$ with $\sum_{i=1}^{N} c_i^0 = 1$ and $z^0 \in H^1(\Omega)$ with $0 \leq z^0 \leq 1$.
Then, there exists a weak solution (c, u, z, w) in the sense of Definition 5.2.3 with $W^{ch} = W^{ch,pol}$. Additionally, $r = -\chi[\partial_z W^{el}(c, \epsilon(u), z)]^+$, where the function $\chi \in L^\infty(\Omega_T)$ satisfies $\chi = 0$ in $\{z > 0\}$ and $0 \leq \chi \leq 1$ in $\{z = 0\}$.

Theorem 5.2.7 (Existence theorem - logarithmic case)
Let the assumptions (5.1) and (5.2) be satisfied and, additionally, let $\Gamma_D = \partial\Omega$ as well as $\Gamma = \gamma\,\mathrm{Id}$ with a constant $\gamma > 0$. Furthermore, let $b \in W^{1,1}(0, T; W^{1,\infty}(\Omega; \mathbb{R}^n))$, $c^0 \in H^1(\Omega; \mathbb{R}^N)$ with $\sum_{i=1}^{N} c_i^0 = 1$, and $c_i^0 > 0$ for all $k = 1, \ldots, N$ and $z^0 \in H^1(\Omega)$ with $0 \leq z^0 \leq 1$ be given.
Then, there exists a weak solution (c, u, z, w) in the sense of Definition 5.2.3 with $W^{ch} = W^{ch,\log}$. Additionally, $c_k > 0$ for all $k = 1, \ldots, N$ and $r = -\chi[\partial_z W^{el}(c, \epsilon(u), z)]^+$, where the function $\chi \in L^\infty(\Omega_T)$ satisfies $\chi = 0$ in $\{z > 0\}$ and $0 \leq \chi \leq 1$ in $\{z = 0\}$.

5.3   Proofs of the existence theorems

5.3.1   Existence of weak solutions for the regularized system

The proof is based on the argumentation of Chapter 4. Arguments similar to Chapter 4 are only sketched below.

Since $\varepsilon > 0$ is fixed in this section, we omit the ε-dependence in the notation, e.g. $\mathcal{E}$ always means here $\mathcal{E}_\varepsilon$ and so on. Furthermore, z^0 is assumed to be in $W^{1,q}(\Omega)$ in this subsection.

Proof of Theorem 5.2.5.

1. *Step: constructing time-discrete solutions.*

   Set u^0 to be a minimizer of $u \mapsto \mathcal{E}(c^0, u, z^0)$ defined on the space $W^{1,4}(\Omega)$ with the constraint $u|_D = b(0)|_D$ in the sense of traces.

Let the closed subspace $\mathfrak{Q}_M^m$ of $H^1(\Omega; \mathbb{R}^N) \times H^1(\Omega; \mathbb{R}^n) \times W^{1,q}(\Omega)$ be defined by:

$$\mathfrak{Q}_M^m = \left\{ \begin{array}{l} c \in H^1(\Omega; \mathbb{R}^N), \\ u \in W^{1,4}(\Omega; \mathbb{R}^n), \\ z \in W^{1,q}(\Omega) \end{array} \middle| \begin{array}{l} \int_\Omega (c - c^0)\, dx = 0 \text{ for C-H systems,} \\ u|_{\Gamma_D} = b(m\tau)|_{\Gamma_D}, \\ 0 \leq z \leq z_M^{m-1}. \end{array} \right\}$$

Based on the initial triple (c^0, u^0, z^0), we construct (c_M^m, u_M^m, z_M^m) for $m = 1, \ldots, M$ recursively by minimizing the following functional $\mathbb{E}_M^m : \mathfrak{Q}_M^m \to \mathbb{R}$:

$$\mathbb{E}_M^m(c, u, z) := \widetilde{\mathcal{E}}(c, u, z) + \tau \widetilde{\mathcal{R}}\left(\frac{z - z_M^{m-1}}{\tau} \right) + \frac{\tau}{2} \left\| \frac{c - c_M^{m-1}}{\tau} \right\|_X^2 + \frac{\varepsilon\tau}{2} \left\| \frac{c - c_M^{m-1}}{\tau} \right\|_{L^2}^2, \tag{5.20}$$

where X denotes the space $\widetilde{V}_0$ (see (4.23b)) with the scalar-product

$$(c_1 \,|\, c_2)_X := \int_\Omega \mathbb{M} \nabla \mathcal{S}^{-1} c_1 \cdot \nabla \mathcal{S}^{-1} c_2 \, dx$$

in the case of Cahn-Hilliard systems and $X = L^2(\Omega; \mathbb{R}^N)$ with the scalar-product

$$(c_1 \,|\, c_2)_X := \int_\Omega \mathbb{M} c_1 \cdot c_2 \, dx$$

in the case of Allen-Cahn systems.

Note that the last regularization term in (5.20) is not necessary for Allen-Cahn equations due to $X = L^2(\Omega; \mathbb{R}^N)$. To use a uniform approach, we consider this term in both systems. By direct methods of calculus of variations, the minimizer

$$(c_M^m, u_M^m, z_M^m) := \underset{(c,u,z) \in \mathfrak{Q}_M^m}{\arg\min} \ \mathbb{E}_M^m(c, u, z)$$

exists, cf. Lemma 4.3.1. Furthermore, we set

$$w_M^m := \begin{cases} -\mathcal{S}^{-1}\left(\frac{c_M^m - c_M^{m-1}}{\tau} \right) + \lambda_M^m & \text{for C-H systems,} \\ -\mathcal{S}^{-1}\left(\frac{c_M^m - c_M^{m-1}}{\tau} \right) & \text{for A-C systems,} \end{cases}$$

with the Lagrange multiplier λ_M^m (associated with the mass constraint for C-H systems) given by

$$\lambda_M^m := \fint_\Omega \mathbb{P}(\partial_c W^{\mathrm{ch,pol}}(c_M^m) + \partial_c W^{\mathrm{el}}(c_M^m, \epsilon(u_M^m), z_M^m)) \, dx.$$

By means of the time incremental solutions $g_M^m := (u_M^m, c_M^m, z_M^m, w_M^m)$, we introduce the piecewise constant interpolations $g_M := (u_M, c_M, z_M, w_M)$, $g_M^- := (u_M^-, c_M^-, z_M^-, w_M^-)$ and the linear interpolation $\widehat{g}_M := (\widehat{u}_M, \widehat{c}_M, \widehat{z}_M, \widehat{w}_M)$ as

$$g_M(t) := g_M^m \qquad\qquad \text{for } t \in ((m-1)\tau, m\tau],$$

$$g_M^-(t) := g_M^m \qquad\qquad \text{for } t \in \big[m\tau, (m+1)\tau\big),$$

$$\widehat{g}_M(t) := \beta g_M^m + (1-\beta)g_M^{m-1} \quad \text{for } t \in \big[(m-1)\tau, m\tau\big) \text{ and } \beta = \frac{t}{\tau} - (m-1)$$

and the piecewise time constants t_M and t_M^- as

$$t_M := \min\{m\tau \,|\, m \in \mathbb{N}_0 \text{ and } m\tau \geq t\},$$
$$t_M^- := \min\{(m-1)\tau \,|\, m \in \mathbb{N}_0 \text{ and } m\tau \geq t\}.$$

Due to the minimization properties of (c_M^m, u_M^m, z_M^m), we establish the following variational formulas and energy estimate (cf. Lemma 4.3.2):

Lemma 5.3.1 (Euler-Lagrange equations and energy inequality)
The functions g_M, g_M^- and $\widehat{g}_M$ satisfy the following properties for all $t \in (0,T)$:

(i) for all $\zeta \in H^1(\Omega; \mathbb{R}^N)$:

$$\int_\Omega (\partial_t \widehat{c}_M(t)) \cdot \zeta \, \mathrm{d}x = -\langle \mathcal{S}w_M(t), \zeta \rangle \tag{5.21}$$

(ii) for all $\zeta \in H^1(\Omega; \mathbb{R}^N)$:

$$\int_\Omega w_M(t) \cdot \zeta \, \mathrm{d}x = \int_\Omega \big(\mathbb{P}\Gamma \nabla c_M(t) : \nabla \zeta + \mathbb{P}\partial_c W^{\mathrm{ch,pol}}(c_M(t)) \cdot \zeta\big) \, \mathrm{d}x$$
$$+ \int_\Omega \big(\mathbb{P}\partial_c W^{\mathrm{el}}(c_M(t), \epsilon(u_M(t)), z_M(t)) \cdot \zeta + \varepsilon \partial_t \widehat{c}_M(t) \cdot \zeta\big) \, \mathrm{d}x \tag{5.22}$$

(iii) for all $\zeta \in W^{1,4}_{\Gamma_D}(\Omega; \mathbb{R}^n)$:

$$\int_\Omega \big(\partial_e W^{\mathrm{el}}(c_M(t), \epsilon(u_M(t)), z_M(t)) : \epsilon(\zeta) + \varepsilon |\nabla u_M(t)|^2 \nabla u_M(t) : \nabla \zeta\big) \, \mathrm{d}x = 0 \tag{5.23}$$

(iv) for all $\zeta \in W^{1,q}(\Omega)$ with $0 \leq \zeta + z_M(t) \leq z_M^-(t)$:

$$\int_\Omega \big((\varepsilon |\nabla z_M(t)|^{q-2} + 1)\nabla z_M(t) \cdot \nabla \zeta + \partial_z W^{\mathrm{el}}(c_M(t), \epsilon(u_M(t)), z_M(t))\zeta\big) \, \mathrm{d}x$$
$$+ \int_\Omega (-\alpha + \beta(\partial_t \widehat{z}_M(t)))\zeta \, \mathrm{d}x \geq 0 \tag{5.24}$$

(v) energy inequality:

$$\mathcal{E}(c_M(t), u_M(t), z_M(t)) + \int_0^{t_M} \int_\Omega \Big(-\alpha \partial_t \widehat{z}_M + \frac{\beta}{2}|\partial_t \widehat{z}_M|^2 + \frac{\varepsilon}{2}|\partial_t \widehat{c}_M|^2\Big) \, \mathrm{d}x \, \mathrm{d}s$$
$$+ \int_0^{t_M} \frac{1}{2} \langle \mathcal{S}w_M(s), w_M(s) \rangle \, \mathrm{d}s - \mathcal{E}(c^0, u^0, z^0)$$

$$\leq \int_0^{t_M} \int_\Omega \partial_e W^{\mathrm{el}}(c_M^-, \epsilon(u_M^- + b - b_M^-), z_M) : e(\partial_t b)\, \mathrm{d}x\, \mathrm{d}s$$

$$+ \varepsilon \int_0^{t_M} \int_\Omega |\nabla u_M^- + \nabla b - \nabla b_M^-|^2 \nabla(u_M^- + b - b_M^-) : \nabla \partial_t b\, \mathrm{d}x\, \mathrm{d}s.$$

$$(5.25)$$

2. Step: identifying convergent subsequences.

The energy estimate (v) in Lemma 5.3.1, growth condition (5.2e) and a Gronwall estimation argument lead to a-priori estimates for the energy $\mathcal{E}(c_M(t), u_M(t), z_M(t))$ and for the terms $\|\partial_t \widehat{z}_M\|_{L^2(\Omega_T)}$, $\|\partial_t \widehat{c}_M\|_{L^2(\Omega_T)}$ and $\int_0^T \langle \mathcal{S} w_M(s), w_M(s) \rangle\, \mathrm{d}s$. By standard compactness arguments and a compactness theorem from Aubin and Lions (see Theorem 2.3.9 (i)), we deduce the following weak convergence properties, cf. Lemma 4.3.3, Corollary 4.3.4 and Lemma 4.3.5:

Lemma 5.3.2 (Weak convergence of the time incremental solutions)
There exists a subsequence $\{M_k\}$ and an element (u, c, z, w) satisfying (i) from Definition 5.2.2 such that:

(i) $u_{M_k} \overset{*}{\rightharpoonup} u$ in $L^\infty(0, T; W^{1,4}(\Omega; \mathbb{R}^n))$,

(ii) $z_{M_k}, z_{M_k}^-, \widehat{z}_{M_k} \overset{*}{\rightharpoonup} z$ in $L^\infty(0, T; W^{1,q}(\Omega))$,
   $z_{M_k}(t), z_{M_k}^-(t), \widehat{z}_{M_k}(t) \rightharpoonup z(t)$ in $W^{1,q}(\Omega)$ for a.e. t,
   $z_{M_k}, z_{M_k}^-, \widehat{z}_{M_k} \to z$ a.e. in Ω_T,
   $\widehat{z}_{M_k} \rightharpoonup z$ in $H^1(0, T; L^2(\Omega))$,

(iii) $c_{M_k}, c_{M_k}^-, \widehat{c}_{M_k} \overset{*}{\rightharpoonup} c$ in $L^\infty(0, T; H^1(\Omega; \mathbb{R}^N))$,
   $c_{M_k}(t), c_{M_k}^-(t), \widehat{c}_{M_k}(t) \rightharpoonup c(t)$ in $H^1(\Omega; \mathbb{R}^N)$ for a.e. t,
   $c_{M_k}, c_{M_k}^-, \widehat{c}_{M_k} \to c$ a.e. in Ω_T,
   $\widehat{c}_{M_k} \rightharpoonup c$ in $H^1(0, T; L^2(\Omega; \mathbb{R}^N))$

(iv) $w_{M_k} \rightharpoonup w$ in $L^2(0, T; H^1(\Omega; \mathbb{R}^N))$ *for C-H systems,*
   $w_{M_k} \rightharpoonup w$ in $L^2(\Omega_T; \mathbb{R}^N)$ *for A-C systems*

as $k \to \infty$.

By using uniform convexity estimates and by exploiting the Euler-Lagrange equations and Lemma 2.3.18, we can even prove stronger convergence properties.

Lemma 5.3.3 (Strong convergence of the time incremental solutions)
There exists a subsequence $\{M_k\}$ such that:

(i) $u_{M_k}, u_{M_k}^- \to u$ in $L^4(0, T; W^{1,4}(\Omega; \mathbb{R}^n))$,
   $u_{M_k}(t), u_{M_k}^-(t) \to u(t)$ in $W^{1,4}(\Omega; \mathbb{R}^n)$ for a.e. t,
   $u_{M_k}, u_{M_k}^- \to u$ a.e. in Ω_T,

(ii) $c_{M_k}, c_{M_k}^-, \widehat{c}_{M_k} \to c$ in $L^{2^*}(0,T; H^1(\Omega; \mathbb{R}^N))$,

$c_{M_k}(t), c_{M_k}^-(t), \widehat{c}_{M_k}(t) \to c(t)$ in $H^1(\Omega; \mathbb{R}^N)$ for a.e. t,

$c_{M_k}, c_{M_k}^-, \widehat{c}_{M_k} \to c$ a.e. in Ω_T,

$\widehat{c}_{M_k} \rightharpoonup c$ in $H^1(0,T; L^2(\Omega; \mathbb{R}^N))$,

(iii) $z_{M_k}, z_{M_k}^-, \widehat{z}_{M_k} \to z$ in $L^q(0,T; W^{1,q}(\Omega))$,

$z_{M_k}(t), z_{M_k}^-(t), \widehat{z}_{M_k}(t) \to z(t)$ in $W^{1,q}(\Omega)$ for a.e. t,

$z_{M_k}, z_{M_k}^-, \widehat{z}_{M_k} \to z$ a.e. in Ω_T,

$\widehat{z}_{M_k} \rightharpoonup z$ in $H^1(0,T; L^2(\Omega))$

as $k \to \infty$.

Proof. We omit the index k in the proof.

(i) We refer to Lemma 4.3.6.

(ii) Weak convergence properties for c_M, c_M^- and $\widehat{c}_{M_k}$ have been proven in Lemma 5.3.2. It remains to show strong convergence of ∇c_M to ∇c in $L^2(\Omega_T; \mathbb{R}^{N \times n})$. Because then, the strong convergences of ∇c_M^- and $\nabla \widehat{c}_M$ to ∇c in $L^2(\Omega_T; \mathbb{R}^{N \times n})$ follow as in Lemma 4.3.7.

By the compact embedding $H^1(\Omega; \mathbb{R}^N) \hookrightarrow L^{2^*/2+1}(\Omega; \mathbb{R}^N)$ and Lemma 5.3.2, we get $\|c_M(t) - c(t)\|_{L^{2^*/2+1}(\Omega; \mathbb{R}^N)} \to 0$ as $M \to \infty$ for a.e. $t \in (0,T)$. The boundedness property ess $\sup_{t \in [0,T]} \|c_M(t) - c(t)\|_{L^{2^*/2+1}(\Omega; \mathbb{R}^N)} < C$ for all $M \in \mathbb{N}$ and Lebesgue's convergence theorem yield $c_M \to c$ as $M \to \infty$ in $L^{2^*/2+1}(\Omega_T; \mathbb{R}^N)$. Testing (5.22) with $\zeta = c_M(t)$ and with $\zeta = c(t)$ gives after integration from $t = 0$ to $t = T$:

$$\int_{\Omega_T} \mathbb{P}\Gamma \nabla c_M : \nabla c_M \, dx \, dt = \int_{\Omega_T} \left(w_M \cdot c_M - \mathbb{P}\partial_c W^{\mathrm{ch,pol}}(c_M) \cdot c_M \right) dx \, dt$$
$$- \int_{\Omega_T} \left(\mathbb{P}\partial_c W^{\mathrm{el}}(c_M, \epsilon(u_M), z_M) \cdot c_M + \varepsilon \partial_t \widehat{c}_M \cdot c_M \right) dx \, dt,$$

$$\int_{\Omega_T} \mathbb{P}\Gamma \nabla c_M : \nabla c \, dx \, dt = \int_{\Omega_T} \left(w_M \cdot c - \mathbb{P}\partial_c W^{\mathrm{ch,pol}}(c_M) \cdot c \right) dx \, dt$$
$$- \int_{\Omega_T} \left(\mathbb{P}\partial_c W^{\mathrm{el}}(c_M, \epsilon(u_M), z_M) \cdot c + \varepsilon \partial_t \widehat{c}_M \cdot c \right) dx \, dt.$$

Passing to $M \to \infty$ and comparing the right sides of the equations shows

$$\int_{\Omega_T} \mathbb{P}\Gamma \nabla c_M : \nabla c_M \, dx \, dt \to \int_{\Omega_T} \mathbb{P}\Gamma \nabla c : \nabla c \, dx \, dt.$$

By using the properties $\mathbb{P}\nabla c_M = \nabla c_M$ and $\mathbb{P}\nabla c = \nabla c$, we eventually obtain

$$\int_{\Omega_T} \Gamma \nabla c_M : \nabla c_M \, dx \, dt \to \int_{\Omega_T} \Gamma \nabla c : \nabla c \, dx \, dt$$

We end up with

$$\int_{\Omega_T} \Gamma(\nabla c_M - \nabla c) : (\nabla c_M - \nabla c) \, dx \, dt \to 0.$$

Therefore, $\nabla c_M \to \nabla c$ in $L^2(\Omega_T; \mathbb{R}^{N \times n})$ since Γ is positive definite.

(iii) Applying Lemma 2.3.18 with $f = z$ and $f_M = z_M^-$ and $\zeta = z$ gives an approximation sequence $\{\zeta_M\} \subseteq L^q(0, T; W_+^{1,q}(\Omega))$ with the properties:

$$\zeta_M \to z \text{ in } L^q(0, T; W^{1,q}(\Omega)), \tag{5.26a}$$
$$0 \le \zeta_M \le z_M^- \text{ for all } M \in \mathbb{N}. \tag{5.26b}$$

The estimate

$$C_{\text{uc}} |\nabla z_M - \nabla z|^q \le (|\nabla z_M|^{q-2} \nabla z_M - |\nabla z|^{q-2} \nabla z) \cdot \nabla(z_M - z)$$

where $C_{\text{uc}} > 0$ is a constant and equation (5.24) tested with $\zeta = \zeta_M(t) - z_M(t)$ (possible due to (5.26b)) yield:

$$C_{\text{uc}} \int_{\Omega_T} \varepsilon |\nabla z_M - \nabla z|^q \, dx \, dt + \int_{\Omega_T} |\nabla z_M - \nabla z|^2 \, dx \, dt$$

$$\le \int_{\Omega_T} \left((\varepsilon |\nabla z_M|^{q-2} + 1) \nabla z_M - (\varepsilon |\nabla z|^{q-2} + 1) \nabla z \right) \cdot \nabla(z_M - z) \, dx \, dt$$

$$= \int_{\Omega_T} (\varepsilon |\nabla z_M|^{q-2} + 1) \nabla z_M \cdot \nabla(z_M - \zeta_M) \, dx \, dt$$

$$+ \int_{\Omega_T} \left((\varepsilon |\nabla z_M|^{q-2} + 1) \nabla z_M \cdot \nabla(\zeta_M - z) - (\varepsilon |\nabla z|^{q-2} + 1) \nabla z \cdot \nabla(z_M - z) \right) dx \, dt$$

$$\le \int_{\Omega_T} (\partial_z W^{\text{el}}(c_M, \epsilon(u_M), z_M) - \alpha + \beta \partial_t \widehat{z}_M)(\zeta_M - z_M) \, dx \, dt$$

$$+ \int_{\Omega_T} \left((\varepsilon |\nabla z_M|^{q-2} + 1) \nabla z_M \cdot \nabla(\zeta_M - z) - (\varepsilon |\nabla z|^{q-2} + 1) \nabla z \cdot \nabla(z_M - z) \right) dx \, dt$$

$$\le \underbrace{\|\partial_z W^{\text{el}}(c_M, \epsilon(u_M), z_M) - \alpha + \beta \partial_t \widehat{z}_M\|_{L^2(\Omega_T)}}_{\text{bounded}} \|\zeta_M - z_M\|_{L^2(\Omega_T)}$$

$$+ \underbrace{(\varepsilon \|\nabla z_M\|_{L^q(\Omega_T)}^{q-1} + \|\nabla z_M\|_{L^{q/(q-1)}(\Omega_T)})}_{\text{bounded}} \|\nabla \zeta_M - \nabla z\|_{L^q(\Omega_T)}$$

$$- \int_{\Omega_T} (\varepsilon |\nabla z|^{q-2} + 1) \nabla z \cdot \nabla(z_M - z) \, dx \, dt$$

Due to (5.26a) and $z_M \overset{\star}{\rightharpoonup} z$ in $L^\infty(0, T; W^{1,q}(\Omega))$ as well as $z_M \to z$ in $L^2(\Omega_T)$, each term on the right hand side converges to 0 as $M \to \infty$.

The strong convergences of ∇z_M^- and $\nabla \widehat{z}_M$ to ∇z in $L^q(\Omega_T; \mathbb{R}^n)$ follow analogously. $\qquad\square$

3. *Step: establishing the precise energy inequality.* In this step we establish an energy inequality which is sharper than the energy inequality in (5.25). Note, that compared to (5.25) the factor $1/2$ in front of $\langle \mathcal{S}w_M(s), w_M(s) \rangle$ is missing. To simplify notation, we omit the index k in the following.

Lemma 5.3.4 *For every $t \in (0, T)$:*

$$\mathcal{E}(c_M(t), u_M(t), z_M(t)) + \int_0^{t_M} \int_\Omega \left(-\alpha \partial_t \widehat{z}_M + \beta |\partial_t \widehat{z}_M|^2 + \varepsilon |\partial_t \widehat{c}_M|^2 \right) \mathrm{d}x \, \mathrm{d}s$$

$$+ \int_0^{t_M} \langle \mathcal{S}w_M(s), w_M(s) \rangle \, \mathrm{d}s - \mathcal{E}(c^0, u^0, z^0)$$

$$\leq \int_0^{t_M} \int_\Omega \partial_e W^{\mathrm{el}}(\epsilon(u_M^- + b - b_M^-), c_M^-, z_M) : e(\partial_t b) \, \mathrm{d}x \, \mathrm{d}s$$

$$+ \varepsilon \int_0^{t_M} \int_\Omega |\nabla u_M^- + \nabla b - \nabla b_M^-|^2 \nabla(u_M^- + b - b_M^-) : \nabla \partial_t b \, \mathrm{d}x \, \mathrm{d}s + \kappa_M$$

with $\kappa_M \to 0$ as $M \to \infty$.

Proof. Applying the estimate $\mathbb{E}_M^m(c_M^m, u_M^m, z_M^m) \leq \mathbb{E}_M^m(c_M^m, u_M^{m-1} + b_M^m - b_M^{m-1}, z_M^m)$ for $m = 1$ to $\frac{t_M}{\tau}$ yields (cf. Lemma 4.3.10):

$$\mathcal{E}(c_M(t), u_M(t), z_M(t)) - \mathcal{E}(c^0, u^0, z^0)$$

$$\leq \varepsilon \int_0^{t_M} \int_\Omega |\nabla(u_M^- + b(s) - b_M^-)|^2 \nabla(u_M^- + b(s) - b_M^-) : \nabla \partial_t b(s) \, \mathrm{d}x \, \mathrm{d}s$$

$$+ \int_0^{t_M} \int_\Omega \partial_e W^{\mathrm{el}}(\epsilon(u_M^- + b - b_M^-), c_M^-, z_M^-) : e(\partial_t b) \, \mathrm{d}x \, \mathrm{d}s$$

$$+ \underbrace{\int_0^{t_M} \int_\Omega \partial_c W^{\mathrm{el}}(\epsilon(u_M^- + b_M - b_M^-), \widehat{c}_M, z_M^-) \cdot \partial_t \widehat{c}_M \, \mathrm{d}x \, \mathrm{d}s}_{(\star)_1}$$

$$+ \underbrace{\int_0^{t_M} \int_\Omega \left(\Gamma \nabla \widehat{c}_M : \nabla \partial_t \widehat{c}_M + \partial_c W^{\mathrm{ch,pol}}(\widehat{c}_M) \cdot \partial_t \widehat{c}_M \right) \mathrm{d}x \, \mathrm{d}s}_{(\star)_2}$$

$$+ \underbrace{\int_0^{t_M} \int_\Omega \partial_z W^{\mathrm{el}}(\epsilon(u_M^- + b_M - b_M^-), c_M, \widehat{z}_M) \partial_t \widehat{z}_M \, \mathrm{d}x \, \mathrm{d}s}_{(\star\star)_1}$$

$$+ \underbrace{\int_0^{t_M} \int_\Omega \left(\varepsilon |\nabla \widehat{z}_M|^{q-2} \nabla \widehat{z}_M \cdot \nabla \partial_t \widehat{z}_M + \nabla \widehat{z}_M \cdot \nabla \partial_t \widehat{z}_M \right) \mathrm{d}x \, \mathrm{d}s}_{(\star\star)_2}.$$

The elementary inequalities

$$(|\nabla \widehat{z}_M|^{q-2} \nabla \widehat{z}_M - |\nabla z_M|^{q-2} \nabla z_M) \cdot \nabla \partial_t \widehat{z}_M \leq 0$$

$$(\nabla \widehat{z}_M - \nabla z_M) \cdot \nabla \partial_t \widehat{z}_M \leq 0$$

and (5.24) tested with $\zeta := -\partial_t \widehat{z}_M(t)\tau$ lead to the estimate:

$$(\star\star)_1 + (\star\star)_2$$
$$\leq -\int_0^{t_M} \int_\Omega \left(-\alpha \partial_t \widehat{z}_M + \beta |\partial_t \widehat{z}_M|^2 \right) dx\, ds$$
$$+ \underbrace{\int_0^{t_M} \int_\Omega (\partial_z W^{\mathrm{el}}(\epsilon(u_M^- + b_M - b_M^-), c_M, \widehat{z}_M) - \partial_z W^{\mathrm{el}}(c_M, \epsilon(u_M), z_M)) \partial_t \widehat{z}_M \, dx\, ds}_{=: \kappa_M^3}.$$

Furthermore,

$$(\star)_1 \leq \int_0^{t_M} \int_\Omega \partial_c W^{\mathrm{el}}(c_M, \epsilon(u_M), z_M) \cdot \partial_t \widehat{c}_M \, dx\, ds$$
$$+ \underbrace{\int_0^{t_M} \int_\Omega (\partial_c W^{\mathrm{el}}(\epsilon(u_M^- + b_M - b_M^-), \widehat{c}_M, z_M^-) - \partial_c W^{\mathrm{el}}(c_M, \epsilon(u_M), z_M)) \cdot \partial_t \widehat{c}_M \, dx\, ds}_{=: \kappa_M^1}.$$

Using the elementary estimate $\Gamma(\nabla \widehat{c}_M - \nabla c_M) : \nabla \partial_t \widehat{c}_M \leq 0$ gives

$$(\star)_2 \leq \int_0^{t_M} \int_\Omega \left(\Gamma \nabla c_M : \nabla \partial_t \widehat{c}_M + \partial_c W^{\mathrm{ch,pol}}(c_M) \cdot \partial_t \widehat{c}_M \right) dx\, ds$$
$$+ \underbrace{\int_0^{t_M} \int_\Omega (\partial_c W^{\mathrm{ch,pol}}(\widehat{c}_M) - \partial_c W^{\mathrm{ch,pol}}(c_M)) \cdot \partial_t \widehat{c}_M \, dx\, ds}_{=: \kappa_M^2}.$$

Hence, applying equations (5.22) with $\zeta = \partial_t \widehat{c}_M(t)$ and (5.21) with $\zeta = w_M(t)$ by noticing $\mathbb{P}\partial_t \widehat{c}_M(t) = \partial_t \widehat{c}_M(t)$ shows

$$(\star)_1 + (\star)_2 \leq -\int_0^{t_M} \langle \mathcal{S}w_M(s), w_M(s) \rangle \, ds - \int_0^{t_M} \int_\Omega \varepsilon |\partial_t \widehat{c}_M|^2 \, dx\, ds + \kappa_M^1 + \kappa_M^2.$$

Lebesgue's generalized convergence theorem, growth conditions (5.2d)-(5.2g) and Lemma 5.3.3 show $\kappa_M := \kappa_M^1 + \kappa_M^2 + \kappa_M^3 \to 0$ as $M \to \infty$. We would like to emphasize that we need the boundedness of ∇u_M in $L^4(\Omega_T; \mathbb{R}^{n \times n})$ and the boundedness of $\partial_t \widehat{c}_M$ in $L^2(\Omega_T; \mathbb{R}^N)$ and $\partial_t \widehat{z}_M$ in $L^2(\Omega_T)$ with respect to M. $\square$

4. *Step: passing to $M \to \infty$.* By using Lemma 5.3.2, Lemma 5.3.3 and equations (5.21), (5.22) and (5.23), we establish (ii), (iii) and (iv) of Definition 5.2.2. Moreover, Lemma 5.3.4 implies

$$\mathcal{E}(c_M(t), u_M(t), z_M(t)) + \int_{\Omega_t} \left(-\alpha \partial_t \widehat{z}_M + \beta |\partial_t \widehat{z}_M|^2 + \varepsilon |\partial_t \widehat{c}_M|^2 \right) dx\, ds$$

$$+ \int_0^t \langle \mathcal{S}w_M(s), w_M(s) \rangle \, \mathrm{d}s - \mathcal{E}(c^0, u^0, z^0)$$

$$\leq \int_0^{t_M} \int_\Omega \partial_e W^{\mathrm{el}}(\epsilon(u_M^- + b - b_M^-), c_M^-, z_M) : e(\partial_t b) \, \mathrm{d}x \, \mathrm{d}s$$

$$+ \varepsilon \int_0^{t_M} \int_\Omega |\nabla u_M^- + \nabla b - \nabla b_M^-|^2 \nabla(u_M^- + b - b_M^-) : \nabla \partial_t b \, \mathrm{d}x \, \mathrm{d}s + \kappa_M.$$

The energy estimate (vi) from Definition 5.2.2 follows from above by using the known convergence properties and weakly semi-continuity arguments.

It remains to show (v) of Definition 5.2.2. We are now able to prove the remaining property.

Lemma 5.3.5 *We have*

$$\int_\Omega \left((\varepsilon |\nabla z(t)|^{q-2} + 1) \nabla z(t) \cdot \nabla \zeta + (\partial_z W^{\mathrm{el}}(c(t), \epsilon(u(t)), z(t)) - \alpha + \beta(\partial_t z(t))) \zeta \right) \mathrm{d}x$$

$$\geq -\langle r(t), \zeta \rangle, \tag{5.27}$$

for all $\zeta \in W_-^{1,q}(\Omega)$ *and for a.e.* $t \in (0, T)$, *where* $r(t) \in L^1(\Omega) \subseteq (W^{1,q}(\Omega))^*$ *is given by*

$$r(t) := -\chi_{\{z(t)=0\}} [\partial_z W^{\mathrm{el}}(c(t), \epsilon(u(t)), z(t))]^+. \tag{5.28}$$

Proof. First of all, we take any test function $\zeta \in L^q(0, T; W_-^{1,q}(\Omega))$ with $\{\zeta = 0\} \supseteq \{z = 0\}$. Lemma 2.3.18 gives a sequence $\{\zeta_M\} \subseteq L^q(0, T; W_-^{1,q}(\Omega))$ with $\zeta_M \to \zeta$ in $L^q(0, T; W_-^{1,q}(\Omega))$ and $0 \geq \nu \zeta_M(t) \geq -z_M(t)$ where ν depends on M and t. Therefore (5.24) holds for $\zeta = \zeta_M(t)$. Integration from 0 to T and passing to $M \to \infty$ gives

$$\int_{\Omega_T} \left((\varepsilon |\nabla z|^{q-2} + 1) \nabla z \cdot \nabla \zeta + (\partial_z W^{\mathrm{el}}(c, \epsilon(u), z) - \alpha + \beta(\partial_t z)) \zeta \right) \mathrm{d}x \, \mathrm{d}t \geq 0.$$

In other words,

$$\int_\Omega \left((\varepsilon |\nabla z(t)|^{q-2} + 1) \nabla z(t) \cdot \nabla \zeta + \partial_z W^{\mathrm{el}}(c(t), \epsilon(u(t)), z(t)) \zeta \right) \mathrm{d}x$$

$$+ \int_\Omega (-\alpha + \beta(\partial_t z(t))) \zeta \, \mathrm{d}x \geq 0$$

holds for every $\zeta \in W_-^{1,q}(\Omega)$ with $\{\zeta = 0\} \supseteq \{z(t) = 0\}$ and a.e. $t \in (0, T)$. To finish the proof, we need to extend the variational inequality to the whole space $W_-^{1,q}(\Omega)$.

Setting $f = (\varepsilon |\nabla z(t)|^{q-2} + 1) \nabla z(t)$ and $g = \partial_z W^{\mathrm{el}}(c(t), \epsilon(u(t)), z(t)) - \alpha + \beta(\partial_t z(t))$, Lemma 2.3.19 shows for every $\zeta \in W_-^{1,q}(\Omega)$

$$\int_\Omega \left((\varepsilon |\nabla z(t)|^{q-2} + 1) \nabla z(t) \cdot \nabla \zeta + (\partial_z W^{\mathrm{el}}(c(t), \epsilon(u(t)), z(t)) - \alpha + \beta(\partial_t z(t))) \zeta \right) \mathrm{d}x$$

$$\geq \int_{\{z(t)=0\}} [\partial_z W^{\mathrm{el}}(c(t), \epsilon(u(t)), z(t)) - \alpha + \beta(\partial_t z(t))]^+ \zeta \, \mathrm{d}x$$

$$\geq \int_{\{z(t)=0\}} [\partial_z W^{\mathrm{el}}(c(t), \epsilon(u(t)), z(t))]^+ \zeta \, \mathrm{d}x.$$

Now, variational inequality (5.27) follows by setting

$$r(t) := -\chi_{\{z(t)=0\}} [\partial_z W^{\mathrm{el}}(c(t), \epsilon(u(t)), z(t))]^+.$$

$\square$

Remark 5.3.6 *Lemma 5.3.5 gives more information than (v) from Definition 5.2.2. It provides a special choice for $r(t)$ given by (5.28).*

Theorem 5.2.5 is now proven. $\square$

5.3.2 Existence of weak solutions for the limit system - polynomial case

In this chapter, we show that an appropriate subsequence of the regularized solutions $(c_\varepsilon, u_\varepsilon, z_\varepsilon, w_\varepsilon)$ for $\varepsilon \in (0,1)$ of Definition 5.2.2 converges in "some sense" to a limit (c, u, z, w) which satisfies the limit equations given in Definition 5.2.3. Since the initial damage profile z^0 is in $H^1(\Omega)$, we approximate z^0 by a sequence $\{z_\varepsilon^0\}$ in $W^{1,q}(\Omega)$ such that $z_\varepsilon^0 \to z^0$ in $H^1(\Omega)$ as $\varepsilon \to 0^+$.

Using the energy inequality and Gronwall's inequality, we establish again the following energy estimate.

Lemma 5.3.7 *We have*

$$\mathcal{E}_\varepsilon(c_\varepsilon(t), u_\varepsilon(t), z_\varepsilon(t)) + \int_0^t \int_\Omega \left(-\alpha \partial_t z_\varepsilon + \beta |\partial_t z_\varepsilon|^2 + \varepsilon |\partial_t c_\varepsilon|^2 \right) \mathrm{d}x \, \mathrm{d}s + \int_0^t \langle \mathcal{S} w_\varepsilon(s), w_\varepsilon(s) \rangle \, \mathrm{d}s$$
$$\leq C(\mathcal{E}_\varepsilon(c^0, u_\varepsilon^0, z_\varepsilon^0) + 1)$$

for a.e. $t \in (0,T)$ and every $\varepsilon \in (0,1)$.

Since $\mathcal{E}_\varepsilon(c^0, u_\varepsilon^0, z_\varepsilon^0) \leq \mathcal{E}_\varepsilon(c^0, u_1^0, z_\varepsilon^0) \leq \mathcal{E}_1(c^0, u_1^0, z_\varepsilon^0)$, the left hand side is also uniformly bounded with respect to a.e. $t \in (0,T)$ and every $\varepsilon \in (0,1)$. By using standard compactness theorems and uniform convexity properties of W^{el} (see (5.2a)), we obtain the following convergence properties (cf. Lemma 4.3.11 and Lemma 4.3.12).

Lemma 5.3.8 (Convergence properties) *There exists a subsequence $\{\varepsilon_k\}$ with $\varepsilon_k \to 0^+$ as $k \to \infty$ and an element (c, u, z, w) satisfying (i) of Definition 5.2.3 such that*

(i) $u_{\varepsilon_k} \to u$ in $L^2(0, T; H^1(\Omega; \mathbb{R}^n))$,
   $\sqrt[3]{\varepsilon_k} \nabla u_{\varepsilon_k} \to 0$ in $L^\infty(0, T; L^4(\Omega; \mathbb{R}^n))$,
   $u_{\varepsilon_k}(t) \to u(t)$ in $H^1(\Omega; \mathbb{R}^n)$ for a.e. t,
   $u_{\varepsilon_k} \to u$ a.e. in Ω_T,
   $u^0_{\varepsilon_k} \to u^0$ in $H^1(\Omega; \mathbb{R}^n)$,
   $\sqrt[3]{\varepsilon_k} \nabla u^0_{\varepsilon_k} \to 0$ in $L^4(\Omega; \mathbb{R}^n)$,

(ii) $c_{\varepsilon_k} \overset{\star}{\rightharpoonup} c$ in $L^\infty(0, T; H^1(\Omega; \mathbb{R}^N))$,
   $\varepsilon_k \partial_t c_{\varepsilon_k} \to 0$ in $L^2(\Omega_T; \mathbb{R}^N)$,
   $c_{\varepsilon_k}(t) \rightharpoonup c(t)$ in $H^1(\Omega; \mathbb{R}^N)$ for a.e. t,
   $c_{\varepsilon_k} \to c$ a.e. in Ω_T,

(iii) $z_{\varepsilon_k} \overset{\star}{\rightharpoonup} z$ in $L^\infty(0, T; H^1(\Omega))$,
   $\sqrt[q-1]{\varepsilon_k} \nabla z_{\varepsilon_k} \to 0$ in $L^\infty(0, T; L^q(\Omega))$,
   $z_{\varepsilon_k}(t) \rightharpoonup z(t)$ in $H^1(\Omega)$ for a.e. t,
   $z_{\varepsilon_k} \to z$ a.e. in Ω_T,
   $z_{\varepsilon_k} \rightharpoonup z$ in $H^1(0, T; L^2(\Omega))$

as $k \to \infty$. We additionally obtain for Cahn-Hilliard systems

$$w_{\varepsilon_k} \rightharpoonup w \text{ in } L^2(0, T; H^1(\Omega; \mathbb{R}^N))$$

and for Allen-Cahn systems

$$w_{\varepsilon_k} \rightharpoonup w \text{ in } L^2(\Omega_T; \mathbb{R}^N),$$
$$c_{\varepsilon_k} \rightharpoonup c \text{ in } H^1(0, T; L^2(\Omega; \mathbb{R}^N))$$

as $k \to \infty$.

As before, we will omit the index k in the subscripts below.

Remark 5.3.9 *We would like to mention that the arguments in Lemma 4.3.7 and Lemma 4.3.8 cannot be adapted to prove strong convergence properties of ∇c_ε and ∇z_ε due to the more generous growth condition (5.2d) as well as the use of Lemma 2.3.18 where the compact embedding $W^{1,q}(\Omega) \hookrightarrow \mathcal{C}^{0,\alpha}(\overline{\Omega})$ for $q > n$ with $\alpha > 0$ and $\alpha < 1 - \frac{n}{q}$ is exploited.*

We are now able to establish existence of weak solutions of (3.27)-(3.28) in the polynomial case $W^{\text{ch}} = W^{\text{ch,pol}}$.

Proof of Theorem 5.2.6. Whenever we refer in the following to (5.8)-(5.13) the functions c, u, z, w and r are substituted by $c_\varepsilon, u_\varepsilon, z_\varepsilon, w_\varepsilon$ and r_ε. Moreover, Lemma 5.3.8 is used without mentioning in the following. It remains to prove (ii)-(vi) from Definition 5.2.3:

(ii) Let $\zeta \in L^2(0, T; H^1(\Omega; \mathbb{R}^N))$ with $\partial_t \zeta \in L^2(\Omega_T; \mathbb{R}^N)$ and $\zeta(T) = 0$. Integration from $t = 0$ to $t = T$ of (5.8) and integration by parts yield

$$\int_{\Omega_T} (c_\varepsilon - c^0) \cdot \partial_t \zeta \, dx \, ds = \int_0^T \langle \mathcal{S}w_\varepsilon, \zeta \rangle \, ds.$$

Passing to $\varepsilon \to 0^+$ shows (ii) of Definition 5.2.3.

(iii) Let $\zeta \in L^2(0, T; H^1(\Omega; \mathbb{R}^N)) \cap L^\infty(\Omega_T; \mathbb{R}^N)$. Integration from $t = 0$ to $t = T$ of (5.9) and passing to $\varepsilon \to 0^+$ yield

$$\int_{\Omega_T} w \cdot \zeta \, dx \, ds = \int_{\Omega_T} \left(\mathbb{P}\Gamma \nabla c : \nabla \zeta + (\mathbb{P}\partial_c W^{\mathrm{ch,pol}}(c) + \mathbb{P}\partial_c W^{\mathrm{el}}(c, \epsilon(u), z)) \cdot \zeta \right) dx \, ds.$$

Note that

$$\left| \int_{\Omega_T} \varepsilon \partial_t c_\varepsilon \cdot \zeta \, dx \, ds \right| \le \varepsilon \| \partial_t c_\varepsilon \|_{L^2(\Omega_T; \mathbb{R}^N)} \| \zeta \|_{L^2(\Omega_T; \mathbb{R}^N)} \to 0$$

as $\varepsilon \to 0^+$. This shows (iii) of Definition 5.2.3 with $\partial_c W^{\mathrm{ch}} = \partial_c W^{\mathrm{ch,pol}}$.

(iv) Let $\zeta \in W^{1,4}_{\Gamma_D}(\Omega; \mathbb{R}^n)$ be arbitrary. Passing to $\varepsilon \to 0^+$ in (5.10) yields for a.e. $t \in (0, T)$

$$\int_\Omega \partial_e W^{\mathrm{el}}(c(t), \epsilon(u(t)), z(t)) : \epsilon(\zeta) \, dx = 0, \tag{5.29}$$

by noticing

$$\left| \int_\Omega \varepsilon |\nabla u_\varepsilon(t)|^2 \nabla u_\varepsilon(t) : \nabla \zeta \, dx \right| \le \varepsilon \| \nabla u_\varepsilon(t) \|_{L^4(\Omega)}^3 \| \zeta \|_{L^4(\Omega)} \to 0.$$

A density argument shows that (5.29) also holds for all $\zeta \in H^1_{\Gamma_D}(\Omega; \mathbb{R}^n)$. Therefore, (iv) of Definition 5.2.3 is shown.

(v) The characteristic functions $\chi_{\{z_\varepsilon = 0\}}$ are bounded in $L^\infty(\Omega_T)$ with respect to $\varepsilon \in (0, 1)$. We select a subsequence such that $\chi_{\{z_{\varepsilon_k} = 0\}} \overset{\star}{\rightharpoonup} \chi$ in $L^\infty(\Omega_T)$ as $k \to \infty$. In the following, we will omit the index k in the notation. Integrating (5.11) from $t = 0$ to $t = T$ and passing to $\varepsilon \to 0^+$ show

$$\int_{\Omega_T} \left(\nabla z \cdot \nabla \zeta + (\partial_z W^{\mathrm{el}}(c, \epsilon(u), z) - \alpha + \beta(\partial_t z)) \zeta \right) dx \, dt$$

$$\ge \int_{\Omega_T} \chi [\partial_z W^{\mathrm{el}}(c, \epsilon(u), z)]^+ \zeta \, dx \, ds \tag{5.30}$$

for all $\zeta \in L^q(0, T; W^{1,q}_-(\Omega)) \cap L^\infty(\Omega_T)$. We also used the fact that

$$\left| \int_{\Omega_T} \varepsilon |\nabla z_\varepsilon|^{q-2} \nabla z_\varepsilon \cdot \nabla \zeta \, dx \, ds \right| \le \varepsilon \| \nabla z_\varepsilon \|_{L^q(\Omega_T)}^{q-1} \| \nabla \zeta \|_{L^q(\Omega_T)} \to 0.$$

It follows that

$$\int_\Omega \left(\nabla z(t) \cdot \nabla \zeta + (\partial_z W^{\mathrm{el}}(c(t), \epsilon(u(t)), z(t)) - \alpha + \beta(\partial_t z(t))) \zeta \right) dx$$

$$\geq \int_\Omega \chi(t)[\partial_z W^{\mathrm{el}}(c(t), \epsilon(u(t)), z(t))]^+ \zeta \, dx$$

for all $\zeta \in H^1_-(\Omega) \cap L^\infty(\Omega)$ and a.e. $t \in (0, T)$. Set $r := -\chi[\partial_z W^{\mathrm{el}}(c, \epsilon(u), z)]^+$. For every $\xi \in L^\infty((0, T))$ with $\xi \geq 0$ a.e. on $(0, T)$ and every $\zeta \in H^1_+(\Omega) \cap L^\infty(\Omega)$ we also have

$$0 \geq \int_0^T \left(\int_\Omega r_\varepsilon(t)(\zeta - z_\varepsilon(t)) \, dx \right) \xi(t) \, dt = \int_{\Omega_T} r_\varepsilon(\zeta - z_\varepsilon)\xi \, dx \, dt$$

$$\to \int_{\Omega_T} r(\zeta - z)\xi \, dx \, dt = \int_0^T \left(\int_\Omega r(t)(\zeta - z(t)) \, dx \right) \xi(t) \, dt.$$

This shows $\int_\Omega r(t)(\zeta - z(t)) \, dx \leq 0$ for a.e. $t \in (0, T)$. Hence, we obtain the inequalities (v) of Definition 5.2.3.

(vi) Weakly semi-continuity arguments lead to

$$\liminf_{\varepsilon \to 0^+} \Big(\mathcal{E}_\varepsilon(c_\varepsilon(t), u_\varepsilon(t), z_\varepsilon(t))$$

$$+ \int_{\Omega_t} \left(\alpha|\partial_t z_\varepsilon| + \beta|\partial_t z_\varepsilon|^2 + \varepsilon|\partial_t c_\varepsilon|^2 \right) dx \, ds + \int_0^t \langle \mathcal{S}w_\varepsilon, w_\varepsilon \rangle \, ds \Big)$$

$$\geq \mathcal{E}(c(t), u(t), z(t)) + \int_{\Omega_t} \left(\alpha|\partial_t z| + \beta|\partial_t z|^2 \right) dx \, ds + \int_0^t \langle \mathcal{S}w, w \rangle \, ds.$$

Testing (5.10) with $\zeta = u^0_\varepsilon - b(0)$ and (iv) of Definition 5.2.3 with $\zeta = u^0 - b(0)$ yield

$$\varepsilon \int_\Omega |\nabla u^0_\varepsilon|^4 \, dx = \varepsilon \int_\Omega |\nabla u^0_\varepsilon|^2 \nabla u^0_\varepsilon : \nabla b(0) \, dx$$

$$- \int_\Omega \partial_e W^{\mathrm{el}}(c^0, \epsilon(u^0_\varepsilon), z^0_\varepsilon) : \epsilon(u^0_\varepsilon - b(0)) \, dx$$

$$\to - \int_\Omega \partial_e W^{\mathrm{el}}(c^0, \epsilon(u^0), z^0) : \epsilon(u^0 - b(0)) \, dx = 0$$

as $\varepsilon \to 0^+$.

Therefore, we can pass to the limit $\varepsilon \to 0^+$ in (5.13) and obtain (vi) from Definition 5.2.3. $\qquad \square$

5.3.3   Higher integrability of the strain tensor

To prove existence results for chemical free energies of logarithmic type, a higher integrability result for the strain tensor based on [Gar00, Gar05b] will be established. We adapt the higher integrability result for solutions of the elliptic equation of the form

$$\begin{cases} \operatorname{div}(\partial_e W^{\mathrm{el}}(c, \epsilon(u))) = 0 & \text{on } \Omega_T, \\ \partial_e W^{\mathrm{el}}(c, \epsilon(u)) \cdot \nu = \sigma^\star \cdot \nu & \text{on } (\partial\Omega)_T \end{cases}$$

to our setting with non-constant Dirichlet boundary data b and the additional damage variable z. In the following, we will use the assumption $\Gamma_{\mathrm{D}} = \partial\Omega$.

Theorem 5.3.10 (Higher integrability) *Let $b \in W^{1,\infty}(\Omega; \mathbb{R}^n)$, $z \in L^\infty(\Omega)$ with $0 \le z \le 1$ a.e. in Ω and $c \in L^\mu(\Omega; \mathbb{R}^N)$ for some $\mu > 4$. Then there exists some $p \in (2, \mu/2]$ such that for all $u \in H^1(\Omega; \mathbb{R}^n)$ which satisfy $u|_{\Gamma_{\mathrm{D}}} = b|_{\Gamma_{\mathrm{D}}}$ and*

$$\int_\Omega \partial_e W^{\mathrm{el}}(c, \epsilon(u), z) : \epsilon(\zeta)\, \mathrm{d}x = 0 \text{ for all } \zeta \in H^1_{\Gamma_{\mathrm{D}}}(\Omega; \mathbb{R}^n), \tag{5.31}$$

we obtain $u \in W^{1,p}(\Omega; \mathbb{R}^n)$ and

$$\|\nabla u\|_{L^p(\Omega; \mathbb{R}^{n\times n})} \le C(\|\nabla u\|_{L^2(\Omega; \mathbb{R}^{n\times n})} + \|c\|^2_{L^{2p}(\Omega; \mathbb{R}^N)} + 1). \tag{5.32}$$

The positive constants p and C are independent of c, u, z.

Proof. The proof is based on [Gar00, Lemma 4.4 and Theorem 4.3] and uses a covering argument. However, due to the non-constant boundary condition, we need to apply a further variant of the Sobolev-Poincaré inequality (see Theorem 2.3.6 (ii)).

(i) *Higher integrability at the boundary.*

Let $x_0 \in \partial\Omega$. Then, there exist an $R_0 > 0$ and a bi-Lipschitz function $\tau : Q \to \mathbb{R}^n$ with the open cube $Q := Q_{R_0}(0)$ such that $x_0 \in \tau(Q)$ and

$$\tau(Q^+) \subseteq \Omega,$$
$$\tau(Q^-) \subseteq \mathbb{R}^n \setminus \overline\Omega,$$

where $Q^+ := \{x \in Q \mid x_n > 0\}$ and $Q^- := \{x \in Q \mid x_n < 0\}$. Define the transformed functions $\tilde{u}, \tilde{b} \in H^1(Q^+; \mathbb{R}^n)$, $\tilde{c} \in H^1(Q^+)$ and $\tilde{z} \in L^\infty(Q^+)$ as

$$(\tilde{u}, \tilde{b}, \tilde{c}, \tilde{z})(x) := (u, b, c, z)(\tau(x)).$$

To proceed, let $y_0 \in Q$ and $R < \frac{1}{2}\mathrm{dist}(y_0, \partial Q)$ and define for each $R' > 0$ the sets

$$Q^\pm_{R'}(y_0) := \{x \in Q_{R'}(y_0) \mid x_n \gtrless 0\}.$$

We distinguish three cases:

Case 1. We first consider the case $Q^+_R(y_0) \ne \emptyset$ and $Q^-_{\frac{3}{2}R}(y_0) \ne \emptyset$.

The bi-Lipschitz continuity of τ ensures

$$\mathrm{dist}(\tau(\partial Q^+_{2R}(y_0)) \cap \Omega, \tau(\partial Q^+_R(y_0)) \cap \Omega) > RC_1,$$

where $C_1 > 0$ is independent of R and y_0. Let $\xi \in \mathcal{C}_0^\infty(\Omega)$ be a cutoff function with the properties:

(a) $\xi = 0$ in $\Omega \setminus \tau(Q_{2R}(y_0))$,

(b) $0 \leq \xi \leq 1$ in Ω,

(c) $\xi \equiv 1$ in $\tau(Q_R(y_0)) \cap \Omega$,

(d) $|\nabla \xi| \leq \frac{2}{C_1} R^{-1}$.

Testing (5.31) with $\zeta = \xi^2(u - b)$, using the computation

$$\epsilon(\zeta) = \xi^2 \epsilon(u) - \xi^2 \epsilon(b) + \xi((u - b)(\nabla \xi)^t + \nabla \xi (u - b)^t),$$

and (5.2b), we obtain

$$\int_\Omega \xi^2 \partial_e W^{\mathrm{el}}(c, \epsilon(u), z) : \epsilon(u)\, \mathrm{d}x$$
$$= \int_\Omega \xi^2 \partial_e W^{\mathrm{el}}(c, \epsilon(u), z) : \epsilon(b)\, \mathrm{d}x - 2\int_\Omega \xi \partial_e W^{\mathrm{el}}(c, \epsilon(u), z) : ((u - b)(\nabla \xi)^t)\, \mathrm{d}x.$$
$$(5.33)$$

By (5.2a), (5.2e) and (5.2c) we also have the estimates

$$\eta|\epsilon(u)|^2 \leq \partial_e W^{\mathrm{el}}(c, \epsilon(u), z) : \epsilon(u) + C(|c|^2 + 1)|\epsilon(u)|,$$
$$|\partial_e W^{\mathrm{el}}(c, \epsilon(u), z) : ((u - b)(\nabla \xi)^t| \leq \frac{C}{R}(|\epsilon(u)| + |c|^2 + 1)|u - b|,$$
$$|\partial_e W^{\mathrm{el}}(c, \epsilon(u), z) : \epsilon(b)| \leq (|\epsilon(u)| + |c|^2 + 1)|\epsilon(b)|.$$

Therefore, (5.33) can be estimated by

$$\eta \int_\Omega \xi^2|\epsilon(u)|^2\, \mathrm{d}x \leq C\int_\Omega \xi^2(|c|^2 + 1)|\epsilon(u)|\, \mathrm{d}x + \frac{C}{R}\int_\Omega \xi(|\epsilon(u)| + |c|^2 + 1)|u - b|\, \mathrm{d}x$$
$$+ C\int_\Omega \xi^2(|\epsilon(u)| + |c|^2 + 1)|\epsilon(b)|\, \mathrm{d}x.$$

Young's inequality yields

$$c_1 \int_\Omega \xi^2|\epsilon(u)|^2\, \mathrm{d}x \leq C\int_\Omega \xi^2(|c|^4 + 1)\, \mathrm{d}x + \frac{C}{R^2}\int_\Omega |u - b|^2\, \mathrm{d}x. \qquad (5.34)$$

We choose $\mu = f_{Q_{2R}^+(y_0)} \tilde{u}\, \mathrm{d}x$. The calculation $e(\xi(u - \mu)) = \xi \epsilon(u) + \frac{1}{2}((u - \mu)(\nabla \xi)^t + \nabla \xi (u - \mu)^t)$ leads to

$$\int_\Omega |e(\xi(u - \mu))|^2\, \mathrm{d}x \leq 2\left(\int_\Omega \xi^2|\epsilon(u)|^2\, \mathrm{d}x + \int_\Omega |u - \mu|^2|\nabla \xi|^2\, \mathrm{d}x\right). \qquad (5.35)$$

Combining (5.34) and (5.35), applying Korn's inequality for H^1-functions with zero boundary values and using (a) and (b) gives

$$\int_\Omega |\nabla(\xi(u - \mu))|^2\, \mathrm{d}x \leq C\int_{\tau(Q_{2R}^+(y_0))} (|c|^4 + 1)\, \mathrm{d}x + \frac{C}{R^2}\int_{\tau(Q_{2R}^+(y_0))} |u - b|^2\, \mathrm{d}x$$

$$+ \frac{C}{R^2} \int_{\tau(Q_{2R}^+(y_0))} |u - \mu|^2 \, \mathrm{d}x.$$

Because of $\nabla(\xi(u-\mu)) = \xi\nabla u + (u-\mu)(\nabla\xi)^t$ we derive by (a) and (c) the following type of Caccioppoli-inequality:

$$\int_{\tau(Q_R^+(y_0))} |\nabla u|^2 \, \mathrm{d}x \leq C \int_{\tau(Q_{2R}^+(y_0))} (|c|^4 + 1) \, \mathrm{d}x + \frac{C}{R^2} \int_{\tau(Q_{2R}^+(y_0))} |u - b|^2 \, \mathrm{d}x$$
$$+ \frac{C}{R^2} \int_{\tau(Q_{2R}^+(y_0))} |u - \mu|^2 \, \mathrm{d}x.$$

Integral transformation by τ implies

$$\int_{Q_R^+(y_0)} |\nabla\widetilde{u}|^2 \, \mathrm{d}x \leq C \int_{Q_{2R}^+(y_0)} (|\widetilde{c}|^4 + 1) \, \mathrm{d}x + \frac{C}{R^2} \int_{Q_{2R}^+(y_0)} |\widetilde{u} - \widetilde{b}|^2 \, \mathrm{d}x$$
$$+ \frac{C}{R^2} \int_{Q_{2R}^+(y_0)} |\widetilde{u} - \mu|^2 \, \mathrm{d}x.$$

The condition $Q_{\frac{3}{2}R}^-(y_0) \neq \emptyset$ and $D = \partial\Omega$ imply that $\widetilde{u}-\widetilde{b}$ vanishes on $\partial(Q_{2R}^+(y_0)) \cap \mathbb{R}^{n-1}\times\{0\}$. Therefore, we obtain by applying both variants of the Sobolev-Poincaré inequality in Theorem 2.3.6 for $p = 2n/(n+2)$:

$$\int_{Q_R^+(y_0)} |\nabla\widetilde{u}|^2 \, \mathrm{d}x \leq C \int_{Q_{2R}^+(y_0)} (|\widetilde{c}|^4 + 1) \, \mathrm{d}x + \frac{C}{R^2} \mathcal{L}^n(Q_{2R}^+(y_0))^{-\frac{2}{n}} \mathrm{diam}(Q_{2R}^+(y_0))^2$$
$$\cdot \left[\left(\int_{Q_{2R}^+(y_0)} |\nabla\widetilde{u} - \nabla\widetilde{b}|^{\frac{2n}{n+2}} \, \mathrm{d}x \right)^{\frac{n+2}{n}} + \left(\int_{Q_{2R}^+(y_0)} |\nabla\widetilde{u}|^{\frac{2n}{n+2}} \, \mathrm{d}x \right)^{\frac{n+2}{n}} \right].$$
$$(5.36)$$

Note that if $n = 1$ we cannot apply Theorem 2.3.6 because of $p = 2n/(n+2) < 1$. In this case, we can work with the inequalities in Theorem 2.3.6 where p is substituted by 1 and $p^\star$ is substituted by 2. However, we will only treat the more delicate case $n \geq 2$ in the following.

The estimates $\mathrm{diam}(Q_{2R}^+(y_0)) \leq CR$ and $\mathcal{L}^n(Q_{2R}^+(y_0)) \geq R^n$ (because of $Q_R^+(y_0) \neq \emptyset$) show

$$\mathcal{L}^n(Q_{2R}^+(y_0))^{-\frac{2}{n}} \mathrm{diam}(Q_{2R}^+(y_0))^2 \leq C. \qquad (5.37)$$

Now, dividing (5.36) by $\mathcal{L}^n(Q_R(y_0))$ and using (5.37) and

$$\frac{1}{R^2} \frac{1}{\mathcal{L}^n(Q_{2R}(y_0))} \leq C \left(\frac{1}{\mathcal{L}^n(Q_{2R}(y_0))} \right)^{\frac{n+2}{n}}$$

gives

$$\frac{1}{\mathcal{L}^n(Q_R(y_0))} \int_{Q_R^+(y_0)} |\nabla\widetilde{u}|^2 \, \mathrm{d}x \leq \frac{C}{\mathcal{L}^n(Q_{2R}(y_0))} \int_{Q_{2R}^+(y_0)} (|\widetilde{c}|^4 + 1) \, \mathrm{d}x$$

$$+ C \left(\frac{1}{\mathcal{L}^n(Q_{2R}(y_0))} \int_{Q_{2R}^+(y_0)} |\nabla \tilde{u}|^{\frac{2n}{n+2}} \, dx \right)^{\frac{n+2}{n}}$$

$$+ C \left(\frac{1}{\mathcal{L}^n(Q_{2R}(y_0))} \int_{Q_{2R}^+(y_0)} |\nabla \tilde{b}|^{\frac{2n}{n+2}} \, dx \right)^{\frac{n+2}{n}}.$$

Observe that

$$\left(\frac{1}{\mathcal{L}^n(Q_{2R}(y_0))} \int_{Q_{2R}^+(y_0)} |\nabla \tilde{b}|^{\frac{2n}{n+2}} \, dx \right)^{\frac{n+2}{n}} \leq \|\nabla b\|_{L^\infty(\Omega)}^2.$$

Define the following functions on Q:

$$g(x) := \begin{cases} |\nabla \tilde{u}(x)|^{\frac{2n}{n+2}} & \text{for } x \in Q^+, \\ 0 & \text{for } x \in Q \setminus Q^+ \end{cases}$$

and

$$f(x) := \begin{cases} C(|\bar{c}|^4 + \|\nabla b\|_{L^\infty(\Omega)}^2 + 1)^{\frac{n}{n+2}} & \text{for } x \in Q^+, \\ 0 & \text{for } x \in Q \setminus Q^+. \end{cases}$$

We eventually get

$$\fint_{Q_R(y_0)} g^{\frac{n+2}{n}} \, dx \leq \fint_{Q_{2R}(y_0)} f^{\frac{n+2}{n}} \, dx + C \left(\fint_{Q_{2R}(y_0)} g \, dx \right)^{\frac{n+2}{n}}. \tag{5.38}$$

Case 2. Assume $Q_R^+(y_0) \neq \emptyset$ and $Q_{\frac{3}{2}R}^-(y_0) = \emptyset$.

The bi-Lipschitz continuity of τ implies

$$\text{dist}(\tau(\partial Q_{\frac{3}{2}R}(y_0)), \tau(\partial Q_R(y_0))) > RC_1,$$

where $C_1 > 0$ is independent of R and y_0. Therefore, we can choose a cutoff function $\xi \in \mathbb{C}_0^\infty(\Omega)$ which satisfies

(a) $\xi = 0$ in $\Omega \setminus \tau(Q_{\frac{3}{2}R}(x_0))$,   (c) $\xi \equiv 1$ in $\tau(Q_R(x_0))$,

(b) $0 \leq \xi \leq 1$ in Ω,   (d) $|\nabla \xi| \leq \frac{2}{C_1} R^{-1}$.

Testing (5.31) with $\xi = \zeta^2(u - \mu)$ and $\mu := \fint_{Q_{\frac{3}{2}R}(x_0)} \tilde{u} \, dx$ yields as in the previous case

$$\int_{\tau(Q_R(x_0))} |\nabla u|^2 \, dx \leq C \int_{\tau(Q_{\frac{3}{2}R}(x_0))} (|c|^4 + 1) \, dx + \frac{C}{R^2} \int_{\tau(Q_{\frac{3}{2}R}(x_0))} |u - \mu|^2 \, dx.$$

Consequently,

$$
\fint_{Q_R(x_0)} |\nabla\widetilde{u}|^2 \, dx \leq C \fint_{Q_{\frac{3}{2}R}(x_0)} (|\widetilde{c}|^4 + 1) \, dx + C \left(\fint_{Q_{\frac{3}{2}R}(x_0)} |\nabla\widetilde{u}|^{\frac{2n}{n+2}} \, dx \right)^{\frac{n+2}{n}}.
$$

Therefore, the inequality (5.38) is also satisfied in this case.

Case 3. Assume $Q_R^+(y_0) = \emptyset$.

In this case, inequality (5.38) trivially holds.

In all three cases, the reverse Hölder inequality (see Theorem 2.3.10) shows $g \in L_{\mathrm{loc}}^s(Q)$ for all $s \in \left[\frac{n+2}{n}, \frac{n+2}{n} + \varepsilon \right)$ and some $\varepsilon > 0$ depending on R_0 and n.

(ii) *Higher integrability in the interior.*

This case follows with much less effort and is only sketched here.

Let $x_0 \in \Omega$ be arbitrary and $R > 0$ such that $Q_{2R}(x_0) \subseteq \Omega$. We take a cutoff function $\xi \in \mathcal{C}_0^\infty(\Omega)$ with

(a) $\xi = 0$ in $\Omega \setminus Q_{2R}(x_0)$,   (c) $\xi \equiv 1$ in $Q_R(x_0)$,

(b) $0 \leq \xi \leq 1$ in Ω,   (d) $|\nabla\xi| \leq \frac{2}{R}$.

Testing (5.31) with $\xi = \zeta^2(u - \mu)$ and $\mu = \fint_{Q_{2R}(x_0)} u \, dx$ yields with the same computation as in the case (i):

$$
\int_{Q_R(x_0)} |\nabla u|^2 \, dx \leq C \int_{Q_{2R}(x_0)} (|c|^4 + 1) \, dx + \frac{C}{R^2} \int_{Q_{2R}(x_0)} |u - \mu|^2 \, dx.
$$

The Poincaré-Sobolev inequality implies

$$
\fint_{Q_R(x_0)} |\nabla u|^2 \, dx \leq C \fint_{Q_{2R}(x_0)} (|c|^4 + 1) \, dx + C \left(\fint_{Q_{2R}(x_0)} |\nabla u|^{\frac{2n}{n+2}} \, dx \right)^{\frac{n+2}{n}}.
$$

Applying Theorem 2.3.10 with $g = |\nabla u|^{\frac{2n}{n+2}}$, $q = \frac{n+2}{n}$ and $f = C(|c|^4 + 1)^{\frac{n}{n+2}}$ finishes the proof. $\qquad\square$

5.3.4 Existence of weak solutions for the limit system - logarithmic case

The challenge here is to establish the integral equation (iii) in Definition 5.2.3 because the derivative of the logarithmic free chemical energy (5.3) becomes singular if one of the c_k's approaches 0. We only sketch the proof in this section since all essential ideas can be found in [Gar00, Gar05b]. We use a regularization method suggested in [EL91] and also used in [Gar00, Gar05b].

The energy gradient tensor is assumed to be of the form $\Gamma = \gamma \, \mathrm{Id}$ with a constant $\gamma > 0$. Define a $\mathcal{C}^2(\mathbb{R}^N)$ regularization with the regularization parameter $\delta > 0$ as

$$W^{\mathrm{ch},\delta}(c) := \theta \sum_{k=1}^{N} \phi^\delta(c^k) + \frac{1}{2} c \cdot Ac,$$

with

$$\phi^\delta(x) := \begin{cases} x \log(x) & \text{for } d \geq \delta, \\ x \log(\delta) - \frac{\delta}{2} + \frac{x^2}{2\delta} & \text{for } x < \delta. \end{cases}$$

Elliott and Luckhaus showed that the regularization $W^{\mathrm{ch},\delta}$ is uniformly bounded from below.

Lemma 5.3.11 (cf. [EL91]) *There exist constants $\delta_0 > 0$ and $C > 0$ such that*

$$W^{\mathrm{ch},\delta}(c) \geq -C \qquad \text{for all } c \in \mathbb{R}^N \text{ with } c_1 + \ldots + c_N = 1, \ \delta \in (0, \delta_0).$$

Proof of Theorem 5.2.7. Let $(c_\delta, u_\delta, z_\delta, w_\delta)$ denote a weak solution in the sense of Definition 5.2.3 with the free chemical energy $W^{\mathrm{ch}} = W^{\mathrm{ch},\delta}$. By applying Lemma 5.3.11 and using Gronwall's inequality in the energy inequality (vi) of Definition 5.2.3, we can show a-priori estimates analogous as in Subsection 5.3.2 except the a-priori estimate of w_δ.

In the Allen-Cahn case, we have $\partial_t c_\delta = -Mw_\delta$ and, consequently, the boundedness of $\partial_t c_\delta$ in $L^2(\Omega; \mathbb{R}^N)$ and $w_\delta \in \{x \in \mathbb{R}^N \mid x_1 + \ldots x_N = 0\}$ pointwise lead to boundedness of w_δ in $L^2(\Omega; \mathbb{R}^N)$.

In the case of Cahn-Hilliard systems, we can use the following lemma.

Lemma 5.3.12 ([Gar00, Lemma 4.3]) *There exists a constant $C > 0$ such that for all $\delta \in (0, \delta_0)$*

$$\int_0^T \left(\fint_\Omega \mathbb{P}\partial_c W^{\mathrm{ch},\delta}(c_\delta(t)) \, \mathrm{d}x \right)^2 \mathrm{d}t < C.$$

The proof of this lemma is similar to [Gar00, Lemma 4.3]. Therefore, we will omit the proof.

This lemma and the integral equation

$$\int_\Omega w_\delta(t) \, \mathrm{d}x = \int_\Omega \left(\mathbb{P}\partial_c W^{\mathrm{ch},\delta}(c_\delta(t)) + \mathbb{P}\partial_c W^{\mathrm{el}}(c_\delta(t), \epsilon(u_\delta(t)), z_\delta(t)) \right) \mathrm{d}x$$

together with the already known boundedness properties shows

$$\int_0^T \left(\fint_\Omega w_\delta(t) \, \mathrm{d}x \right)^2 \mathrm{d}t < C$$

for a constant $C > 0$. Therefore w_δ is bounded in $L^2(0, T; H^1(\Omega))$ by Poincaré's inequality. In conclusion, we can extract a subsequence $\{(c_{\delta_k}, u_{\delta_k}, z_{\delta_k}, w_{\delta_k})\}$ such that we have the same convergence properties as in Lemma 5.3.8. As before, we will omit the subscript k.

The remaining crucial step is to show that the limit c satisfies $c^k > 0$ a.e. in Ω_T for all $k = 1, \ldots, N$ and $\partial_c W^{\mathrm{ch}, \delta}(c_\delta) \to \partial_c W^{\mathrm{ch}, \log}(c)$ in $L^1(\Omega_T)$ as $\varepsilon \to 0^+$.

To this end, we need an additional boundedness property.

Lemma 5.3.13 *There exists constants $q > 1$ and $C > 0$ such that for all $\delta \in (0, \delta_0)$ and all $k = 1, \ldots, N$:*

$$\|(\phi^\delta)'(c_\delta^k)\|_{L^q(\Omega_T)} < C.$$

We omit the proof of this lemma, since by utilizing Theorem 5.3.10 the arguments are analogous to [Gar00, Lemma 4.5].

Note that

$$\lim_{\delta \to 0^+} (\phi^\delta)'(c_\delta^k) = \begin{cases} \log(c^k) + 1 & \text{if } \lim_{\delta \to 0^+} c_\delta^k = c^k > 0, \\ \infty & \text{otherwise} \end{cases}$$

holds pointwise a.e. in Ω_T and for all $k = 1, \ldots, N$. Together with Lemma 5.3.13, we obtain

$$c^k > 0 \text{ a.e. in } \Omega_T$$

and

$$(\phi^\delta)'(c_\delta^k) \to \log(c^k) + 1 \text{ a.e. in } \Omega_T.$$

This and Lemma 5.3.13 further shows

$$(\phi^\delta)'(c_\delta^k) \to \log(c^k) + 1 \text{ in } L^1(\Omega_T)$$

by Vitali's convergence theorem (see Proposition 1.19 in [Alt99]). Finally, we can pass to $\delta \to 0^+$ in the equation

$$\int_{\Omega_T} w_\delta \cdot \zeta \, dx \, dt = \int_{\Omega_T} \left(\gamma \nabla c_\delta : \nabla \zeta + \mathbb{P} \partial_c W^{\mathrm{ch}, \delta}(c_\delta) \cdot \zeta + \mathbb{P} \partial_c W^{\mathrm{el}}(\epsilon(u_\delta), c_\delta, z_\delta) \cdot \zeta \right) dx \, dt$$

and obtain (iii) from Definition 5.2.3.

The remaining properties can be easily established as in Section 4. Hence, Theorem 5.2.7 is proven. $\qquad\square$

CHAPTER 6

Complete damage processes

In the preceding chapters, Cahn-Hilliard equations have been coupled with incomplete damage processes. The uniform convexity assumptions in (4.2a) as well as in (5.2a), respectively, prevents the PDE system (3.27) from degeneration (in the elastic energy). However, for a more precise description of damage phenomena, the elastic energy should be allowed to degenerate on maximally damaged regions. Studying this case requires further mathematical tools such as Γ-convergence of regularized free energies, representation of shrinking sets with Lipschitz domains and space-time local Sobolev spaces.

In this chapter, we are investigating purely mechanical systems in quasi-static equilibrium undergoing complete damage. See Definition 3.3.1 for a classical description. We will prove local-in-time existence of weak solutions and global-in-time existence of solutions in a weaker sense.

The results and proofs in this chapter can also be found in WIAS preprint no. 1722, see [HK12a].

6.1 Assumptions

In this chapter, we consider the free energy density function ψ in (3.15) and the dissipation potential density function ϕ in (1.4). We assume $f = 0$, $p > n$ and a degenerating elastic energy density W^{el} of the form

$$W^{\mathrm{el}}(e, z) = \frac{1}{2} g(z) \mathbf{C} e : e \tag{6.1}$$

with a positive definite stiffness tensor $\mathbf{C} \in \mathcal{L}(\mathbb{R}^{n \times n}_{\mathrm{sym}})$ satisfying (3.5) and a function $g \in \mathcal{C}^1([0,1]; \mathbb{R}_+)$ with the properties

$$\eta \le g'(z), \tag{6.2a}$$
$$g(0) = 0 \tag{6.2b}$$

for all $z \in [0,1]$ and some constant $\eta > 0$.

Note that complete damage is possible if and only if $g(0) = 0$. The case $g(0) > 0$ would describe incomplete damage processes which is already covered in the mathematical literature (see Chapter 4 and Chapter 5).

It should be remarked that the existence proofs in this chapter also work for potential density function $f \in C^1(\mathbb{R}; \mathbb{R}_+)$ in (1.3) (see Chapter 7).

6.2 Weak formulations and existence results

For the analytical treatment of complete damage systems, we adopt a regularization scheme, where a regularized elastic energy density W^{el}_δ, $\delta > 0$, is used instead of W^{el} in the first instance. More precisely, W^{el}_δ is defined by

$$W^{\mathrm{el}}_\delta(e, z) = \frac{1}{2}(g(z) + \delta)\mathbf{C} e : e. \tag{6.3}$$

In contrast to [MR06] and related works for rate-independent complete damage models, we will not use a purely energetic approach but rather a mixed variational/energetic formulation as presented in Chapter 4 and Chapter 5.

Let $e \in L^2(\Omega; \mathbb{R}^{n \times n}_{\mathrm{sym}})$ and $z \in W^{1,p}(\Omega)$ be given. The associated free energy of the system from Definition 3.3.1 is given by

$$\mathcal{E}(e, z) := \int_\Omega \left(\frac{1}{p} |\nabla z|^p + W^{\mathrm{el}}(e, z) + I_{[0,\infty)}(z) \right) \mathrm{d}x,$$

whereas its δ-regularization with $\delta > 0$ (for later use) is defined as (see (6.3))

$$\mathcal{E}_\delta(e, z) := \int_\Omega \left(\frac{1}{p} |\nabla z|^p + W^{\mathrm{el}}_\delta(e, z) + I_{[0,\infty)}(z) \right) \mathrm{d}x.$$

If e is only defined on a measurable subset $H \subset \Omega$, i.e., $e \in L^2(H; \mathbb{R}^{n \times n}_{\mathrm{sym}})$, we use the convention $\mathcal{E}(e, z) := \mathcal{E}(\widetilde{e}, z)$, where $\widetilde{e} := e$ in H and $\widetilde{e} := 0$ in $\Omega \setminus H$. As in the previous chapters, $\widetilde{\mathcal{E}}$ denotes the energy functional $\mathcal{E}$ without the indicator function.

Note that in contrast to (4.5) and (5.5), the free energy functional here depends on the strain e and the damage variable z.

We are now able to give a weak formulation of the system in an SBV setting (with respect to the damage variable). In accordance to Definition 3.3.1, z is extended on whole $\overline{\Omega_T}$ and when viewed as an $SBV^2(0,T;L^2(\Omega))$-function has a jump at time t if and only if a material exclusion occurs at t.

Definition 6.2.1 (Weak solution for the system (3.23)-(3.24)**)**
A pair (u,z) is called a weak solution of the system given in Definition 3.3.1 with the initial-boundary data (z^0,b) if

(i) *Regularity:*

$$z \in L^\infty(0,T;W^{1,p}(\Omega)) \cap SBV^2(0,T;L^2(\Omega)), \quad u \in L_t^2 H_{x,\mathrm{loc}}^1(F;\mathbb{R}^n)$$

with $\epsilon(u) =: e \in L^2(F;\mathbb{R}_{\mathrm{sym}}^{n\times n})$ where $F := \mathfrak{A}_{\Gamma_D}(\{z^- > 0\}) \subseteq \overline{\Omega_T}$ is a shrinking set. (Note that z^- denotes the limit from the left side w.r.t. the time variable of the BV function z; see Section 2.2.)

(ii) *Quasi-static mechanical equilibrium:*

$$0 = \int_{F(t)} \partial_e W^{\mathrm{el}}(e(t),z(t)) : \epsilon(\zeta)\,\mathrm{d}x \tag{6.4}$$

for a.e. $t \in (0,T)$ and for all $\zeta \in H_{\Gamma_D}^1(\Omega;\mathbb{R}^n)$. Furthermore, $u = b$ on $(\Gamma_D)_T \cap F$.

(iii) *Damage variational inequality:*

$$\int_{F(t)} \left(|\nabla z(t)|^{p-2}\nabla z(t)\cdot\nabla\zeta + \partial_z W^{\mathrm{el}}(e(t),z(t))\zeta\right) \geq \int_\Omega (\alpha - \beta\partial_t^a z(t))\zeta \tag{6.5}$$

$$0 \leq z(t) \text{ in } \Omega,$$
$$0 \geq \partial_t^a z(t) \text{ a.e. in } \Omega$$

for a.e. $t \in (0,T)$ and for all $\zeta \in W^{1,p}(\Omega)$ with $\zeta \leq 0$. The initial value is given by $z^+(0) = z^0$ with $0 \leq z^0 \leq 1$ in Ω.

(iv) *Damage jump condition:*

$$z^+(t) = z^-(t)\mathbb{1}_{F(t)} \text{ in } \Omega \tag{6.6}$$

for all $t \in [0,T]$.

(v) *Weak energy inequality:*

$$\mathcal{E}(e(t),z(t)) + \int_0^t \int_{F(s)} \left(\alpha|\partial_t^a z| + \beta|\partial_t^a z|^2\right)\mathrm{d}x\,\mathrm{d}s + \sum_{s\in J_z\cap(0,t]} \mathcal{J}_s$$

$$\leq \mathfrak{e}_0^+ + \int_0^t \int_{F(s)} \partial_e W^{\text{el}}(e, z) : \epsilon(\partial_t b) \, dx \, ds \tag{6.7}$$

for a.e. $t \in (0, T)$, where the jump part $\mathfrak{J}_s$ satisfies $0 \leq \mathfrak{J}_s$ and is given by

$$\mathfrak{J}_s := \lim_{\tau \to s^-} \operatorname*{ess\,inf}_{\vartheta \in (\tau, s)} \mathcal{E}(e(\vartheta), z(\vartheta)) - \mathfrak{e}_s^+ \tag{6.8}$$

and the values $\mathfrak{e}_s^+ \geq 0$ satisfy the upper energy estimate

$$\mathfrak{e}_s^+ \leq \mathcal{E}(\epsilon(b(s) + \zeta), z^+(s)) \tag{6.9}$$

for all $\zeta \in H^1_{\Gamma_D \cap F(s)}(F(s); \mathbb{R}^n)$.

Remark 6.2.2   (i) For the definition and the properties of the vector-valued Banach space $SBV^2(0, T; L^2(\Omega))$ and the space-time local Sobolev space $L_t^2 H^1_{x, \text{loc}}(F; \mathbb{R}^N)$ for a shrinking set F, we refer to Section 2.2 and Section 2.4. Recall that given $z \in SBV^2(0, T; L^2(\Omega))$ the function $\partial_t^a z$ denotes the absolutely continuous part of the time-derivative of z.

(ii) Lemma 6.3.5 ensures that we have $z^-(t) \mathbb{1}_{F(t)} \in W^{1,p}(\Omega)$ for all times t (see jump condition (6.6)).

(iii) Jump condition (6.6) and the definition of F in Definition 6.2.1 (i) imply $\{z^+(t) > 0\} = F(t)$ for all $t \in [0, T]$. We get

$$\mathcal{E}(e(t), z(t)) = \int_{F(t)} \left(\frac{1}{p} |\nabla z(t)|^p + W^{\text{el}}(e(t), z(t)) \right) dx,$$

which equals $\int_{\{z(t) > 0\}} \left(\frac{1}{p} |\nabla z(t)|^p + W^{\text{el}}(e(t), z(t)) \right) dx$ for a.e. $t \in (0, T)$.

(iv) The jump term $\mathfrak{J}_s$ equals the energy of the excluded material parts at time point s, i.e., $\mathfrak{J}_s = \mathcal{E}(s^-) - \mathcal{E}(s^+)$ (for smooth solutions on F), where $\mathcal{E}(t) := \mathcal{E}(e(t), z(t))$ denotes the energy function along the trajectory. However, for less regular weak solutions as in Definition 6.2.1, the one-sided limits $\mathcal{E}(s^-)$ and $\mathcal{E}(s^+)$ possibly do not exist. But, in any case, $\lim_{\tau \to s^-} \operatorname{ess\,inf}_{\vartheta \in (\tau, s)} \mathcal{E}(\vartheta)$ clearly exists and coincides with $\mathcal{E}(s^-)$ for smooth solutions. The value $\mathcal{E}(s^+)$, on the other hand, can be characterized in a rather indirect way by using upper energy estimates. More precisely, it turns out that $\mathcal{E}(s^+)$ can be substituted by values (denoted by $\mathfrak{e}_s^+$) merely satisfying (6.9). Together with equations (6.4)-(6.7), $\mathfrak{e}_s^+$ is forced to coincide with $\mathcal{E}(s^+)$ for smooth solutions. This is particularly shown in the proof of the following theorem.

Theorem 6.2.3 Let (u, z) be a weak solution according to Definition 6.2.1. We assume the regularity properties $u \in \mathcal{C}^2(\overline{\Omega_T}; \mathbb{R}^n)$ with $u = b$ on $(\Gamma_D)_T$ and $z = \tilde{z}$ in F for a $\tilde{z} \in \mathcal{C}^2(\overline{\Omega_T}; \mathbb{R})$. Then, $(u, \tilde{z})$ is a classical solution according to Definition 3.3.1.

Proof.

We are going to prove the differential inclusion in Definition 3.3.1. The remaining properties follow with much less effort.

The jump condition (6.6) and the regularity assumptions yield for a.e. $(x, t) \in \Omega_T$

$$
\partial_t^a z(x, t) = \begin{cases} \partial_t z(x, t) & \text{if } (x, t) \in F, \\ 0 & \text{if } (x, t) \in \Omega_T \setminus F, \end{cases}
$$

where $\partial_t z(x, t)$ is the classical time-derivative of z at (x, t). In the following, we will make use of this property. First, observe that by the regularity assumptions $q := (e, z) \in SBV(0, T; X)$ with $X := L^2(\Omega; \mathbb{R}^{n \times n}) \times W^{1,p}(\Omega)$.

Applying the chain rule (see Corollary 2.2.6) for the continuously Fréchet-differentiable energy functional $\widetilde{\mathcal{E}}$ and the X-valued SBV function q shows that $\widetilde{\mathcal{E}} \circ q$ is an SBV function and

$$
\begin{aligned}
\mathcal{E}(q(t^+)) - \mathcal{E}(q(0^+)) &= \mathrm{d}(\widetilde{\mathcal{E}} \circ q)((0, t]) \\
&= \int_0^t \left(\langle \mathrm{d}_e \widetilde{\mathcal{E}}(q(s)), \partial_t e(s) \rangle + \langle \mathrm{d}_z \widetilde{\mathcal{E}}(q(s)), \partial_t^a z(s) \rangle \right) \mathrm{d}s \\
&\quad + \sum_{s \in J_z \cap (0, t]} \left(\mathcal{E}(q(s^+)) - \mathcal{E}(q(s^-)) \right).
\end{aligned}
$$

The two terms in the integral on the right hand side can be treated as follows.

- Taking into account $z = 0$ in $\Omega_T \setminus F$ and testing (6.4) with $\zeta = \partial_t u(s) - \partial_t b(s)$, we obtain

$$
\begin{aligned}
\langle \mathrm{d}_e \widetilde{\mathcal{E}}(q(s)), \partial_t e(s) \rangle &= \int_\Omega \partial_e W^{\mathrm{el}}(\epsilon(u(s)), z(s)) : \epsilon(\partial_t u(s)) \, \mathrm{d}x \\
&= \int_{F(s)} \partial_e W^{\mathrm{el}}(\epsilon(u(s)), z(s)) : \epsilon(\partial_t u(s)) \, \mathrm{d}x \\
&= \int_{F(s)} \partial_e W^{\mathrm{el}}(\epsilon(u(s)), z(s)) : \epsilon(\partial_t b(s)) \, \mathrm{d}x.
\end{aligned}
$$

- Using the property $\partial_t^a z = 0$ in $\Omega_T \setminus F$, we obtain

$$
\begin{aligned}
&\langle \mathrm{d}_z \widetilde{\mathcal{E}}(q(s)), \partial_t^a z(s) \rangle \\
&= \int_\Omega \left(|\nabla z(s)|^{p-2} \nabla z(s) \cdot \nabla \partial_t^a z(s) + \partial_z W^{\mathrm{el}}(\epsilon(u(s)), z(s)) \partial_t^a z(s) \right) \mathrm{d}x \\
&= \int_{F(s)} \left(|\nabla z(s)|^{p-2} \nabla z(s) \cdot \nabla \partial_t z(s) + \partial_z W^{\mathrm{el}}(\epsilon(u(s)), z(s)) \partial_t z(s) \right) \mathrm{d}x.
\end{aligned}
$$

Putting the pieces together, we end up with

$$
\mathcal{E}(q(t^+)) + \sum_{s \in J_z \cap (0, t]} \left(\mathcal{E}(q(s^-)) - \mathcal{E}(q(s^+)) \right)
$$

$$= \mathcal{E}(q(0^+)) + \int_0^t \int_{F(s)} \partial_e W^{\mathrm{el}}(\epsilon(u), z) : \epsilon(\partial_t b) \, \mathrm{d}x \, \mathrm{d}s$$

$$+ \int_0^t \int_{F(s)} \left(|\nabla z|^{p-2} \nabla z \cdot \nabla \partial_t z + \partial_z W^{\mathrm{el}}(\epsilon(u), z) \partial_t z\right) \mathrm{d}x \, \mathrm{d}s. \tag{6.10}$$

Note that we have $\mathcal{E}(q(0^+)) = \mathfrak{e}_0^+$. Indeed, passing $t \to 0^+$ in (6.7) yields $\mathcal{E}(q(0^+)) \leq \mathfrak{e}_0^+$. The "$\geq$"-inequality follows from (6.9) tested with $\zeta = u(0) - b(0)$.

Therefore, (6.7) particularly implies

$$\mathcal{E}(q(t^+)) + \int_0^t \int_{F(s)} \left(\alpha|\partial_t z| + \beta|\partial_t z|^2\right) \mathrm{d}x \, \mathrm{d}s + \sum_{s \in J_z \cap (0,t]} \mathfrak{J}_s$$

$$\leq \mathcal{E}(q(0^+)) + \int_0^t \int_{F(s)} \partial_e W^{\mathrm{el}}(e, z) : \epsilon(\partial_t b) \, \mathrm{d}x \, \mathrm{d}s \tag{6.11}$$

Integrating (6.5) on $[0, t]$ with respect to time, testing it with $\zeta = \partial_t^{\mathrm{a}} z \leq 0$, applying it to (6.10) and comparing the result with the energy inequality (6.11) shows

$$\mathcal{E}(q(t^+)) + \sum_{s \in J_z \cap (0,t]} \left(\mathcal{E}(q(s^-)) - \mathcal{E}(q(s^+))\right) + \int_0^t \int_{F(s)} \left(-\alpha\partial_t z + \beta|\partial_t z|^2\right) \mathrm{d}x \, \mathrm{d}s$$

$$\geq \mathcal{E}(q(0^+)) + \int_0^t \int_{F(s)} \partial_e W^{\mathrm{el}}(\epsilon(u), z) : \epsilon(\partial_t b) \, \mathrm{d}x \, \mathrm{d}s$$

$$\geq \mathcal{E}(q(t^+)) + \sum_{s \in J_z \cap (0,t]} \mathfrak{J}_s + \int_0^t \int_{F(s)} \left(-\alpha\partial_t z + \beta|\partial_t z|^2\right) \mathrm{d}x \, \mathrm{d}s. \tag{6.12}$$

Taking also (6.8) into account and using $\mathcal{E}(q(s^-)) = \lim_{\tau \to s^-} \operatorname{ess\,inf}_{\vartheta \in (\tau, s)} \mathcal{E}(q(\vartheta))$, estimate (6.12) yields

$$\sum_{s \in J_z} \mathcal{E}(q(s^+)) \leq \sum_{s \in J_z} \mathfrak{e}_s^+. \tag{6.13}$$

On the other hand, by (6.9), we find $\mathfrak{e}_s^+ \leq \mathcal{E}(q(s^+))$ for all $s \in J_z$. Combining this with (6.13) shows $\mathcal{E}(q(s^+)) = \mathfrak{e}_s^+$ for all $s \in J_z$.

Therefore, $\mathfrak{J}_s = \mathcal{E}(q(s^-)) - \mathcal{E}(q(s^+))$ and (6.12) becomes an equality. Taking also (6.10) into account gives

$$0 = \int_0^t \int_{F(s)} \left(|\nabla z|^{p-2} \nabla z \cdot \nabla \partial_t z + \partial_z W^{\mathrm{el}}(\epsilon(u), z)\partial_t z - \alpha\partial_t z + \beta|\partial_t z|^2\right) \mathrm{d}x \, \mathrm{d}s.$$

Together with the variational inequality (6.5) and the regularity assumptions, we obtain

$$0 \leq \int_{F(s)} \left(-\operatorname{div}(|\nabla z(s)|^{p-2} \nabla z(s)) + \partial_z W^{\mathrm{el}}(\epsilon(u(s)), z(s)) - \alpha + \beta\partial_t z(s)\right)(\zeta - \partial_t z(s)) \, \mathrm{d}x$$

for a.e. $s \in (0,T)$ and for all $\zeta \in L^1(F(s))$ with $\zeta \leq 0$. This leads to

$$0 \leq \Big(-\operatorname{div}(|\nabla z|^{p-2}\nabla z) + \partial_z W^{\mathrm{el}}(\epsilon(u), z) - \alpha + \beta \partial_t z \Big)(\zeta - \partial_t z)$$

for a.e. $(x,t) \in F$. By the regularity assumptions, this inequality holds pointwise in F. Therefore, the differential inclusion in Definition 3.3.1 (ii) is shown. $\qquad\square$

One of the main goals in this work is to prove existence of weak solutions according to Definition 6.2.1. Due to the application of Zorn's lemma used in the global existence proof, analytical problems arise when infinitely many exclusions of material parts occur in arbitrary short time intervals in the "future", i.e., cluster points from the right of the jump set $J_{z^\star}$ (denoted by $C_{z^\star}$ in the following) where $z^\star \in SBV(0,T;L^2(\Omega))$ is given by $z^\star(t) := z(t)\mathbb{1}_{\mathfrak{A}_{\Gamma_{\mathrm{D}}}(\{z^-(t)>0\})}$. See Figure 6.1 for an example. In this case,

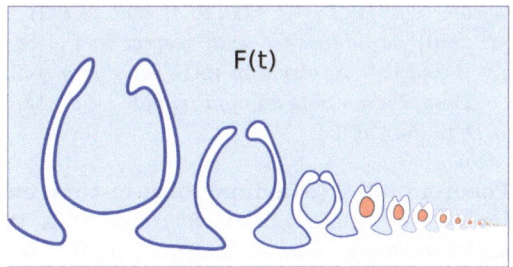

Figure 6.1: *An example of a shrinking set where infinitely many exclusions during an arbitrary small time-interval have been occurred.*

we are only able to prove that the shrinking set F is approximately given by $\mathfrak{A}_{\Gamma_{\mathrm{D}}}(\{z^- > 0\})$ whereas the strain e can still be represented as the symmetric gradient of u in $\mathfrak{A}_{\Gamma_{\mathrm{D}}}(F)$. To be precise, we introduce the following notion.

Definition 6.2.4 (Approximate weak solution for the system (3.23)-(3.24))
A triple (e, u, z) and a shrinking set $F \subseteq \overline{\Omega_T}$ is called an approximate weak solution with fineness $\eta > 0$ of the system according to Definition 3.3.1 with the initial-boundary data (z^0, b) if

(i) *Regularity:*

$$z \in L^\infty(0,T;W^{1,p}(\Omega)) \cap SBV^2(0,T;L^2(\Omega)), \quad u \in L_t^2 H_{x,\mathrm{loc}}^1(\mathfrak{A}_{\Gamma_{\mathrm{D}}}(F);\mathbb{R}^n),$$
$$e \in L^2(F;\mathbb{R}^{n\times n}_{\mathrm{sym}})$$

with $e = \epsilon(u)$ in $\mathfrak{A}_{\Gamma_{\mathrm{D}}}(F)$.

(ii) *Shrinking set properties:*

$$F(t) \supseteq \mathfrak{A}_{\Gamma_{\mathrm{D}}}(\{z^-(t) > 0\}) \text{ for all } t \in [0,T],$$
$$F(t) = \mathfrak{A}_{\Gamma_{\mathrm{D}}}(\{z^-(t) > 0\}) \text{ for all } t \in [0,T] \setminus \bigcup_{t\in C_{z^\star}} [t, t+\eta),$$
$$\mathcal{L}^n\big(F(t) \setminus \mathfrak{A}_{\Gamma_{\mathrm{D}}}(\{z^-(t) > 0\})\big) < \eta \text{ for all } t \in \bigcup_{t\in C_{z^\star}} [t, t+\eta).$$

(iii) Evolutionary equations:

Properties *(ii)-(v)* of Definition 6.2.1 are satisfied.

Remark 6.2.5 *If an approximate weak solution* (e, u, z) *on* F *according to Definition 6.2.4 satisfies* $C_{z^\star} = \emptyset$ *then* (u, z) *is a weak solution according to Definition 6.2.1.*

Theorem 6.2.6 (Global-in-time existence of approximate weak solutions)
Let $b \in W^{1,1}(0, T; W^{1,\infty}(\Omega; \mathbb{R}^n))$ *and* $z^0 \in W^{1,p}(\Omega)$ *with* $0 \leq z^0 \leq 1$ *in* Ω *and the set* $\{z^0 > 0\}$ *be admissible with respect to* Γ_D *be initial-boundary data. Furthermore, let* $\eta > 0$ *and* W^{el} *be given by* (6.1) *satisfying* (6.2).

Then, there exists an approximate weak solution (e, u, z) *with fineness* $\eta > 0$ *according to Definition 6.2.4.*

Theorem 6.2.7 (Maximal local-in-time existence of weak solutions)
Let $b \in W^{1,1}(0, T; W^{1,\infty}(\Omega; \mathbb{R}^n))$ *and* $z^0 \in W^{1,p}(\Omega)$ *with* $0 < \kappa \leq z^0 \leq 1$ *in* Ω *be initial-boundary data. Furthermore, let* W^{el} *be given by* (6.1) *satisfying* (6.2).

Then, there exists a maximal value $\widehat{T} > 0$ *with* $\widehat{T} \leq T$ *and functions* u *and* z *defined on the time interval* $[0, \widehat{T}]$ *such that* (u, z) *is a weak solution according to Definition 6.2.1. Therefore, if* $\widehat{T} < T$, (u, z) *cannot be extended to a weak solution on* $[0, \widehat{T} + \varepsilon]$ *for any* $\varepsilon > 0$.

6.3   Proofs of the existence theorems

6.3.1   Γ-limit of the regularized energy

The construction of the values $\mathfrak{e}_s^+$ in (6.7) satisfying the lower energy bound (6.9) is based on Γ-convergence techniques which will be introduced below. We refer to [BMR09] for the utilization of Γ-convergence in the context with rate-independent complete damage models. In the following, we choose a sequence $\delta_k \to 0^+$ as $k \to \infty$ for the limit passage. For notational convenience, we omit the subscript k.

Definition 6.3.1 (Γ-limit of the δ-regularized reduced energy)
Let $\mathfrak{E}_\delta : H^1(\Omega; \mathbb{R}^n) \times W_{\mathrm{w}}^{1,p}(\Omega) \to \mathbb{R} \cup \{+\infty\}$ *be for* $\delta \geq 0$ *the (regularized) reduced free energy defined by*

$$\mathfrak{E}_\delta(\xi, z) := \inf_{\zeta \in H^1_{\Gamma_D}(\Omega; \mathbb{R}^n)} \mathcal{E}_\delta(\epsilon(\xi + \zeta), z).$$

Then, we denote by $\mathfrak{E}$ *the Γ-limit of* $\mathfrak{E}_\delta$ *as* $\delta \to 0^+$ *with respect to the topology in* $H^1(\Omega; \mathbb{R}^n) \times W_{\mathrm{w}}^{1,p}(\Omega)$. *Here,* $W_{\mathrm{w}}^{1,p}(\Omega)$ *denotes the space* $W^{1,p}(\Omega)$ *with its weak topology.*

Remark 6.3.2 *The existence of the Γ-limit above is ensured because* $\{\mathfrak{E}_\delta\}$ *is non-negative and monotonically decreasing as* $\delta \to 0^+$. *Furthermore,* $\mathfrak{E}$ *is the lower semi-continuous envelope of* $\mathfrak{E}_0$ *in the* $H^1(\Omega; \mathbb{R}^n) \times W_{\mathrm{w}}^{1,p}(\Omega)$ *topology (see Remark 2.3.3).*

To prove properties of the Γ-limit $\mathfrak{E}$ which are needed in Section 6.3.3, we will establish explicit recovery sequences. The proof relies on a substitution method which is introduced in the following.

Define the auxiliary functional $\mathcal{F}_\delta$ by

$$\mathcal{F}_\delta(e, z) := \int_\Omega \left(W_\delta^{\mathrm{el}}(e, z) + I_{[0,\infty)}(z) \right) \mathrm{d}x$$

and assume that $u \in H^1(\Omega; \mathbb{R}^n)$ minimizes $\mathcal{F}_\delta(\epsilon(\cdot), z)$ for given Dirichlet data ξ on Γ_D and damage profile z. Then, by expressing the elastic energy density W_δ^{el} in terms of its derivative $\partial_e W_\delta^{\mathrm{el}}$, i.e., $W_\delta^{\mathrm{el}} = \frac{1}{2}\partial_e W_\delta^{\mathrm{el}} : e$, and by testing the corresponding Euler-Lagrange equation with $\zeta = u - \widetilde{u}$ for a function $\widetilde{u} \in H^1(\Omega; \mathbb{R}^n)$ with $\widetilde{u} = \xi$ on Γ_D, the elastic energy can be rewritten as

$$\int_\Omega W_\delta^{\mathrm{el}}(\epsilon(u), z) \, \mathrm{d}x = \int_\Omega \frac{1}{2}(g(z) + \delta)\mathbf{C}\epsilon(u) : \epsilon(\widetilde{u}) \, \mathrm{d}x. \tag{6.14}$$

Lemma 6.3.3 *For every $\xi \in H^1(\Omega; \mathbb{R}^n)$ and $z \in W_+^{1,p}(\Omega)$ there exists a sequence $\lambda_\delta \to 0^+$ such that $(\xi, [z - \lambda_\delta]^+) \to (\xi, z)$ is a recovery sequence for $\mathcal{F}_\delta \xrightarrow{\Gamma} \mathfrak{F}$ as $\delta \to 0^+$ where $\mathfrak{F}$ is the Γ-limit of $\mathcal{F}_\delta : H^1(\Omega; \mathbb{R}^n) \times W_w^{1,p}(\Omega) \to \mathbb{R} \cup \{+\infty\}$ given by*

$$\mathfrak{F}_\delta(\xi, z) := \min_{\zeta \in H_{\Gamma_D}^1(\Omega; \mathbb{R}^n)} \mathcal{F}_\delta(\epsilon(\xi + \zeta), z)$$

in the $H^1(\Omega; \mathbb{R}^n) \times W_w^{1,p}(\Omega)$ topology.

Proof. The Γ-limit $\mathfrak{F}$ exists by the same argument as in Remark 6.3.2. Let $(\xi_\delta, z_\delta) \to (\xi, z)$ be a recovery sequence. Since $\mathfrak{F}_\delta(\xi_\delta, z_\delta) = \infty$ if z_δ is *not* a non-negative function, we assume WLOG $z_\delta \geq 0$ in $\overline{\Omega}$.

Due to the compact embedding $W^{1,p}(\Omega) \hookrightarrow \mathcal{C}(\overline{\Omega})$, we know $z_\delta \to z$ in $\mathcal{C}(\overline{\Omega})$ as $\delta \to 0^+$. Let $\lambda > 0$ be arbitrary. Then, there exists a $\delta_0 > 0$ such that $\|z - z_\delta\|_{\mathcal{C}(\overline{\Omega})} < \lambda$ for all $0 < \delta < \delta_0$. Noticing $z_\delta, z \geq 0$ in $\overline{\Omega}$, this implies $[z - \lambda]^+ < z_\delta$.

Therefore, we can choose a sequence $\lambda_\delta \to 0^+$ such that $[z - \lambda_\delta]^+ \leq z_\delta$. Note that $[z - \lambda_\delta]^+ \in W^{1,p}(\Omega)$.

Consider the arrangement

$$\mathfrak{F}_\delta(\xi, [z - \lambda_\delta]^+) - \mathfrak{F}_\delta(\xi_\delta, z_\delta) = \underbrace{\mathfrak{F}_\delta(\xi, [z - \lambda_\delta]^+) - \mathfrak{F}_\delta(\xi, z_\delta)}_{A_\delta} + \underbrace{\mathfrak{F}_\delta(\xi, z_\delta) - \mathfrak{F}_\delta(\xi_\delta, z_\delta)}_{B_\delta}.$$

We observe that $A_\delta \leq 0$ because of (note that $[z - \lambda_\delta]^+ \leq z_\delta$)

$$\mathcal{F}_\delta(\epsilon(\xi + \zeta), [z - \lambda_\delta]^+) \leq \mathcal{F}_\delta(\epsilon(\xi + \zeta), z_\delta)$$

for all $\zeta \in H_{\Gamma_D}^1(\Omega; \mathbb{R}^n)$. Let $u_\delta, v_\delta \in H_{\Gamma_D}^1(\Omega; \mathbb{R}^n)$ be given by

$$u_\delta = \operatorname*{arg\,min}_{\zeta \in H_{\Gamma_D}^1(\Omega; \mathbb{R}^n)} \mathcal{F}_\delta(\epsilon(\xi + \zeta), z_\delta),$$

$$v_\delta = \operatorname*{arg\,min}_{\zeta \in H_{\Gamma_D}^1(\Omega; \mathbb{R}^n)} \mathcal{F}_\delta(\epsilon(\xi_\delta + \zeta), z_\delta).$$

Applying the substitution method (see (6.14)) for (u_δ, z_δ) with $\tilde{u} = v_\delta$ and for (v_δ, z_δ) with $\tilde{u} = u_\delta$, we obtain a calculation as follows:

$$B_\delta = \mathcal{F}_\delta(\epsilon(\xi + u_\delta), z_\delta) - \mathcal{F}_\delta(\epsilon(\xi_\delta + v_\delta), z_\delta)$$

$$= \int_\Omega \left(\frac{1}{2}(g(z_\delta) + \delta) \mathbf{C}\epsilon(\xi + u_\delta) : \epsilon(\xi + v_\delta) - \frac{1}{2}(g(z_\delta) + \delta) \mathbf{C}\epsilon(\xi_\delta + v_\delta) : \epsilon(\xi_\delta + u_\delta) \right) dx$$

$$\leq \int_\Omega \frac{1}{2}(g(z_\delta) + \delta) \Big(\mathbf{C}\epsilon(\xi) : \epsilon(\xi) - \mathbf{C}\epsilon(\xi_\delta) : \epsilon(\xi_\delta) \Big) dx$$

$$+ \| \frac{1}{2}(g(z_\delta) + \delta) \mathbf{C}\epsilon(u_\delta + v_\delta) \|_{L^2(\Omega)} \| \epsilon(\xi - \xi_\delta) \|_{L^2(\Omega)}.$$

Using $\xi_\delta \to \xi$ in $H^1(\Omega; \mathbb{R}^n)$, $z_\delta \rightharpoonup z$ in $W^{1,p}(\Omega)$ and the boundedness of $\mathcal{F}_\delta(\epsilon(\xi + u_\delta), z_\delta)$ and $\mathcal{F}_\delta(\epsilon(\xi_\delta + v_\delta), z_\delta)$ with respect to δ, we end up with $\limsup_{\delta \to 0+} B_\delta \leq 0$. Consequently, taking also into account that $(\xi_\delta, z_\delta) \to (\xi, z)$ is a recovery sequence, we obtain

$$\limsup_{\delta \to 0+} \mathcal{F}_\delta(\xi, [z - \lambda_\delta]^+) \leq \limsup_{\delta \to 0+} \mathcal{F}_\delta(\xi_\delta, z_\delta) + \limsup_{\delta \to 0+} A_\delta + \limsup_{\delta \to 0+} B_\delta \leq \mathfrak{F}(\xi, z).$$

$\square$

Corollary 6.3.4   *(i) For every $\xi \in H^1(\Omega; \mathbb{R}^n)$ and $z \in W^{1,p}(\Omega)$*

$$\mathfrak{E}(\xi, z) = \int_\Omega \frac{1}{p} |\nabla z|^p \, dx + \mathfrak{F}(\xi, z).$$

(ii) The recovery sequence $(\xi, [z - \lambda_\delta]^+) \to (\xi, z)$ for $\mathcal{F}_\delta \xrightarrow{\Gamma} \mathfrak{F}$ constructed in Lemma 6.3.3 is a recovery sequence for $\mathfrak{E}_\delta \xrightarrow{\Gamma} \mathfrak{E}$ as well.

(iii) Let $\xi \in H^1(\Omega; \mathbb{R}^n)$, $z \in W^{1,p}(\Omega)$ and $F \subseteq \Omega$ be open such that $\mathbb{1}_F z \in W^{1,p}(\Omega)$. Then $\mathfrak{E}(\xi, \mathbb{1}_F z) \leq \mathfrak{E}(\xi, z)$.

Proof.

(i) Let $(\xi_\delta, z_\delta) \to (\xi, z)$ be a recovery sequence for $\mathfrak{E}_\delta \xrightarrow{\Gamma} \mathfrak{E}$. Hence, $\xi_\delta \to \xi$ in $H^1(\Omega; \mathbb{R}^n)$ and $z_\delta \rightharpoonup z$ in $W^{1,p}(\Omega)$. Applying "$\liminf_{\delta \to 0+}$" on each side of the identity

$$\mathfrak{E}_\delta(\xi_\delta, z_\delta) = \int_\Omega \frac{1}{p} |\nabla z_\delta|^p \, dx + \mathcal{F}_\delta(\xi_\delta, z_\delta) \tag{6.15}$$

yields for a subsequence

$$\mathfrak{E}(\xi, z) \geq \int_\Omega \frac{1}{p} |\nabla z|^p \, dx + \mathfrak{F}(\xi, z).$$

The "$\leq$" - part can be shown by considering a recovery sequence $(\xi, [z - \delta]^+) \to (\xi, z)$ for $\mathcal{F}_\delta \xrightarrow{\Gamma} \mathfrak{F}$ according to Lemma 6.3.3 and applying "$\liminf_{\delta \to 0+}$" in (6.15) with $(\xi_\delta, z_\delta) = (\xi, [z - \delta]^+)$ on both sides.

(ii) This follows from (i).

(iii) Without loss of generality, we assume $0 \leq z$ in Ω. Let $(\xi, [z - \lambda_\delta]^+) \to (\xi, z)$ be a recovery sequence for $\mathfrak{E}_\delta \xrightarrow{\Gamma} \mathfrak{E}$ as in (ii). By assumption, $\mathbb{1}_F[z - \lambda_\delta]^+ \in W^{1,p}(\Omega)$ and $\mathbb{1}_F[z - \lambda_\delta]^+ \to \mathbb{1}_F z$ in $W^{1,p}(\Omega)$ as $\delta \to 0^+$.

Since $\mathcal{E}_\delta(\epsilon(\xi + \zeta), \mathbb{1}_F[z - \lambda_\delta]^+) \leq \mathcal{E}_\delta(\epsilon(\xi + \zeta), [z - \lambda_\delta]^+)$ for all $\zeta \in H^1_{\Gamma_D}(\Omega; \mathbb{R}^n)$, we obtain

$$\inf_{\zeta \in H^1_{\Gamma_D}(\Omega;\mathbb{R}^n)} \mathcal{E}_\delta(\epsilon(\xi + \zeta), \mathbb{1}_F[z - \lambda_\delta]^+) \leq \inf_{\zeta \in H^1_{\Gamma_D}(\Omega;\mathbb{R}^n)} \mathcal{E}_\delta(\epsilon(\xi + \zeta), [z - \lambda_\delta]^+).$$

Therefore,

$$\mathfrak{E}_\delta(\xi, \mathbb{1}_F[z - \lambda_\delta]^+) \leq \mathfrak{E}_\delta(\xi, [z - \lambda_\delta]^+).$$

Passing to $\delta \to 0^+$ yields the claim. $\qquad\qquad\qquad\qquad\qquad\qquad\square$

In combination with Corollary 6.3.4, we need the subsequent truncation property of Sobolev functions.

Lemma 6.3.5 *Let $D, \Omega \subseteq \mathbb{R}^n$ be open sets and $p > n$. Furthermore, assume that a function $f \in W^{1,p}(\Omega)$ fulfills $f = 0$ on $\partial D \setminus \partial \Omega$ (f is here considered as a continuous function due to the embedding $W^{1,p}(\Omega) \hookrightarrow \mathcal{C}(\overline{\Omega})$). Then, $f\mathbb{1}_{\Gamma_D} \in W^{1,p}(\Omega)$.*

Proof. We can reduce the problem to one space dimension by using the following slicing result from [AFP00, Proposition 3.105] for functions $u \in L^p(\Omega)$:

$$u \in W^{1,p}(\Omega) \iff \forall \nu \in \mathbb{S}^{n-1} : u^\nu_x \in W^{1,p}(\Omega^\nu_x) \text{ for } \mathcal{L}^{n-1}\text{-a.e. } x \in \Omega_\nu$$
$$\text{and } \int_{\Omega_\nu} \int_{\Omega^\nu_x} |\nabla u^\nu_x|^p \, dt \, dy < \infty, \tag{6.16}$$

where Ω_ν is the orthogonal projection of Ω to the hyperplane orthogonal to ν and $\Omega^\nu_x := \{t \in \mathbb{R} \,|\, x + t\nu \in \Omega\}$ as well as $u^\nu_x(t) := u(x + t\nu)$.

Applying this result to f, we obtain $f^\nu_x \in W^{1,p}(\Omega^\nu_x)$ for $\mathcal{L}^{n-1}$-a.e. $x \in \Omega_\nu$ and all $\nu \in \mathbb{S}^{n-1}$. Moreover, slices for the function $g := f\mathbb{1}_{\Gamma_D}$ are given by the equation

$$g^\nu_x = f^\nu_x \mathbb{1}_{D^\nu_x}.$$

The function f^ν_x is absolutely continuous. We claim that this is also the case for g^ν_x. To proceed, let $\delta > 0$ be an arbitrary real. Then, we get some constant $\lambda > 0$ such that

$$(a_k, b_k), \ k \in I, \text{ with } a_k \leq b_k \text{ are finitely}$$
$$\text{many disjoint intervals of } \Omega^\nu_x \text{ with } \sum_{k \in I} |a_k - b_k| < \lambda$$
$$\implies \sum_{k \in I} |f^\nu_x(a_k) - f^\nu_x(b_k)| < \delta. \tag{6.17}$$

The property (6.17) is also satisfied for g_x^ν. Indeed, let (a_k, b_k), $k \in I$, with $a_k \leq b_k$ be finitely many disjoint intervals of Ω_x^ν with $\sum_{k \in I} |a_k - b_k| < \lambda$. We define the values $\widetilde{a}_k$ and $\widetilde{b}_k$ in the following way:

$$(\widetilde{a}_k, \widetilde{b}_k) := \begin{cases} (a_k, b_k) & \text{if } a_k, b_k \in D_x^\nu \text{ or } a_k, b_k \notin D_x^\nu, \\ (z, b_k) \text{ for an arbitrary fixed } z \in [a_k, b_k] \cap \partial D_x^\nu & \text{if } a_k \notin D_x^\nu \text{ and } b_k \in D_x^\nu, \\ (a_k, z) \text{ for an arbitrary fixed } z \in [a_k, b_k] \cap \partial D_x^\nu & \text{if } a_k \in D_x^\nu \text{ and } b_k \notin D_x^\nu. \end{cases}$$

We conclude $\sum_{k \in I} |\widetilde{a}_k - \widetilde{b}_k| \leq \sum_{k \in I} |a_k - b_k| \leq \lambda$ and therefore $\sum_{k \in I} |f_x^\nu(\widetilde{a}_k) - f_x^\nu(\widetilde{b}_k)| < \delta$ by (6.17). Taking

$$\sum_{k \in I} |g_x^\nu(a_k) - g_x^\nu(b_k)| = \sum_{k \in I} |g_x^\nu(\widetilde{a}_k) - g_x^\nu(\widetilde{b}_k)| \leq \sum_{k \in I} |f_x^\nu(\widetilde{a}_k) - f_x^\nu(\widetilde{b}_k)|$$

into account, shows that g_x^ν is absolutely continuous and we find $g_x^\nu \in W^{1,p}(\Omega_x^\nu)$.

Moreover, $\int_{\Omega_\nu} \int_{\Omega_x^\nu} |\nabla g_x^\nu|^p \, dt \, dy = \int_{D_\nu} \int_{D_x^\nu} |\nabla f_x^\nu|^p \, dt \, dy < \infty$. Applying (6.16) yields $g \in W^{1,p}(\Omega)$. $\square$

Lemma 6.3.6 *Let $\xi \in H^1(\Omega; \mathbb{R}^n)$ and $z \in W^{1,p}(\Omega)$ with $z \geq 0$. Furthermore, let $u \in H_{\text{loc}}^1(\{z > 0\}; \mathbb{R}^n)$ and for every Lipschitz domain $U \subset\subset \{z > 0\}$, $u = \xi$ on $\Gamma_D \cap \partial U$ in the sense of traces. Then,*

$$\mathfrak{E}(\xi, z) \leq \mathcal{E}(\epsilon(u), z).$$

Proof. Consider an arbitrary $\delta > 0$ and define $z_\delta := [z - \delta]^+$. Since $z \in \mathcal{C}(\overline{\Omega})$, it holds the compact inclusion $\{z_\delta > 0\} \subset\subset \{z > 0\}$. There exists an open set U with Lipschitz boundary such that $\{z_\delta > 0\} \subseteq \overline{U} \subseteq \{z > 0\}$ (e.g. construction of $\partial U \setminus \partial \Omega$ by polygons such that ∂U fulfills the Lipschitz boundary condition; cf. Figure 2.3).

Now, we have $u|_U \in H^1(U; \mathbb{R}^n)$ as well as $u = \xi$ on $\partial U \cap \Gamma_D$. There exists an extension $u_\delta \in H^1(\Omega; \mathbb{R}^n)$ with $u_\delta|_U = u|_U$ and $u_\delta = \xi$ on Γ_D. The monotonicity of $\{\mathfrak{E}_\delta\}$ with respect to δ implies that $\mathfrak{E}$ is the lower semi-continuous envelope of $\widetilde{\mathfrak{E}}(\xi, z) := \inf_{\delta > 0} \mathfrak{E}_\delta(\xi, z)$ in the $H^1(\Omega; \mathbb{R}^n) \times W_w^{1,p}(\Omega)$-topology (see Remark 2.3.3 (ii)). By switching the infima, it holds

$$\widetilde{\mathfrak{E}}(\xi, z) = \begin{cases} \inf_{\zeta \in H_{\Gamma_D}^1(\Omega; \mathbb{R}^n)} \mathcal{E}(\epsilon(\xi + \zeta), z) & \text{if } 0 \leq z \leq 1, \\ \infty & \text{else.} \end{cases}$$

Since $u = u_\delta$ on $\{z_\delta > 0\}$, we get

$$\begin{aligned} \mathfrak{E}(\xi, z) &= \inf_{\xi_\delta \to \xi \text{ in } H^1(\Omega; \mathbb{R}^n)} \inf_{\eta_\delta \to z \text{ in } W^{1,p}(\Omega)} \liminf_{\delta \to 0} \widetilde{\mathfrak{E}}(\xi_\delta, \eta_\delta) \\ &\leq \liminf_{\delta \to 0} \widetilde{\mathfrak{E}}(\xi, z_\delta) \leq \liminf_{\delta \to 0} \mathcal{E}(\epsilon(u_\delta), z_\delta) \\ &\leq \liminf_{\delta \to 0} \mathcal{E}(\epsilon(u), z_\delta) = \mathcal{E}(\epsilon(u), z). \end{aligned}$$

$\square$

6.3.2   Degenerate limit of the regularized system

In the first step of the proofs of Theorem 6.2.6 and Theorem 6.2.7, an existence result for a simplified problem, where no exclusions of material parts are considered, will be shown. The statement we are going to prove in this subsection is given as follows.

Proposition 6.3.7 (Degenerate limit) *Let $b \in W^{1,1}(0, T; W^{1,\infty}(\Omega; \mathbb{R}^n))$ and $z^0 \in W^{1,p}(\Omega)$ with $0 \leq z^0 \leq 1$ be initial-boundary data and let W^{el} be given by (6.1) satisfying (6.2).*
Then, there exist functions

$$z \in L^\infty(0, T; W^{1,p}(\Omega)) \cap H^1(0, T; L^2(\Omega)), \quad u \in L_t^2 H_{x,\mathrm{loc}}^1(\mathfrak{A}_{\Gamma_\mathrm{D}}(\{z > 0\}); \mathbb{R}^n),$$
$$e \in L^2(\{z > 0\}; \mathbb{R}_{\mathrm{sym}}^{n \times n})$$

with $e = \epsilon(u)$ in $\mathfrak{A}_{\Gamma_\mathrm{D}}(\{z > 0\})$ such that the properties (ii)-(v) of Definition 6.2.1 are fulfilled for $F := \{z > 0\}$. Moreover, $\mathfrak{e}_0^+$ (see energy inequality (6.7)) can be chosen to be $\mathfrak{E}(b^0, z^0)$ which satisfies (6.9) by Lemma 6.3.6.

Remark 6.3.8 *Let us consider the degenerate limit functions e, u and z obtained above. Note that we do not know that $F = \{z > 0\}$ equals $\mathfrak{A}_{\Gamma_\mathrm{D}}(\{z > 0\})$ and, if $F \setminus \mathfrak{A}_{\Gamma_\mathrm{D}}(\{z > 0\}) \neq \emptyset$, it is not clear whether u can be extended to a vector-function on F such that $e = \epsilon(u)$ holds in F. On the other hand, we would like to stress that $(u, z^\star)$ with the truncated function $z^\star := z\mathbb{1}_{\mathfrak{A}_{\Gamma_\mathrm{D}}(\{z>0\})}$ also does not necessarily form a weak solution in the sense of Definition 6.2.1. Because $z^\star$ viewed as an $SBV^2(0, T; L^2(\Omega))$ function may have jumps which need to be accounted for in the energy inequality (6.7). The construction of weak solutions will be performed in Section 6.3.3.*

Let $(b^0, z_\delta^0) \to (b^0, z^0)$ with $z_\delta^0 := [z - \lambda_\delta]^+$ and $b^0 := b(0)$ be a recovery sequence of $\mathfrak{E}_\delta \overset{\Gamma}{\to} \mathfrak{E}$ according to Lemma 6.3.4 (ii). A modification of the proof of Theorem 4.2.7 yields the following result.

Theorem 6.3.9 (δ-regularized problem - incomplete damage) *Let $\delta > 0$. For the given initial-boundary data $z_\delta^0 \in W^{1,p}(\Omega)$ and $b \in W^{1,1}(0, T; W^{1,\infty}(\Omega; \mathbb{R}^n))$ there exists a pair $q_\delta = (u_\delta, z_\delta)$ with the subsequent properties:*

(i) The functions are in the following spaces:

$$z_\delta \in L^\infty(0, T; W^{1,p}(\Omega)) \cap H^1(0, T; L^2(\Omega)), \quad u_\delta \in L^\infty(0, T; H^1(\Omega; \mathbb{R}^n)).$$

(ii) Quasi-static mechanical equilibrium:

$$\int_\Omega \partial_e W_\delta^{\mathrm{el}}(\epsilon(u_\delta(t)), z_\delta(t)) : \epsilon(\zeta) \, \mathrm{d}x = 0 \tag{6.18}$$

for a.e. $t \in (0, T)$ and for all $\zeta \in H_{\Gamma_\mathrm{D}}^1(\Omega; \mathbb{R}^n)$. Furthermore, $u_\delta = b$ on the boundary $(\Gamma_\mathrm{D})_T$.

(iii) Damage variational inequality:

$$\int_\Omega \left(|\nabla z_\delta(t)|^{p-2}\nabla z_\delta(t)\cdot\nabla\zeta + \partial_z W_\delta^{\mathrm{el}}(\epsilon(u_\delta(t)), z_\delta(t))\zeta\right) \geq \int_\Omega (\alpha - \beta\partial_t z_\delta(t) - r_\delta(t))\zeta, \tag{6.19}$$

$$z_\delta(t) \geq 0 \ in \ \Omega,$$
$$\partial_t z_\delta(t) \leq 0 \ a.e. \ in \ \Omega$$

for a.e. $t \in (0,T)$ and for all $\zeta \in W_-^{1,p}(\Omega)$ where $r_\delta \in L^1(\Omega_T)$ satisfies

$$\int_\Omega r_\delta(t)(\xi - z_\delta(t))\,\mathrm{d}x \leq 0$$

for a.e. $t \in (0,T)$ and for all $\xi \in W_+^{1,p}(\Omega)$.
The initial value is given by $z_\delta(t=0) = z_\delta^0$ in $\overline{\Omega}$.

(iv) Energy inequality:

$$\mathcal{E}_\delta(\epsilon(u_\delta(t)), z_\delta(t)) + \int_{\Omega_t} \left(\alpha|\partial_t z_\delta| + \beta|\partial_t z_\delta|^2\right)\mathrm{d}x\,\mathrm{d}s$$

$$\leq \mathcal{E}_\delta(\epsilon(u_\delta^0), z_\delta^0) + \int_{\Omega_t} \partial_e W_\delta^{\mathrm{el}}(\epsilon(u_\delta), z_\delta) : \epsilon(\partial_t b)\,\mathrm{d}x\,\mathrm{d}s \tag{6.20}$$

holds for a.e. $t \in (0,T)$ where u_δ^0 minimizes $\mathcal{E}_\delta(\epsilon(\cdot), z_\delta^0)$ in $H^1(\Omega;\mathbb{R}^n)$ with Dirichlet data b^0 on Γ_{D}.

Moreover, r_δ in (iv) can be chosen to be

$$r_\delta = -\chi_\delta \partial_z W^{\mathrm{el}}(\epsilon(u_\delta), z_\delta) \tag{6.21}$$

with $\chi_\delta \in L^\infty(\Omega)$ fulfilling $\chi_\delta = 0$ on $\{z_\delta > 0\}$ and $0 \leq \chi_\delta \leq 1$ on $\{z_\delta = 0\}$.

We consider a sequence $\{\delta_M\}_{M\in\mathbb{N}} \subseteq (0,1)$ with $\delta_M \to 0^+$ as $M \to \infty$ and for every $M \in \mathbb{N}$ a weak solution $(u_{\delta_M}, z_{\delta_M})$ of the incomplete damage problem according to Theorem 6.3.9. The index M is omitted in the following. We agree that $e_\delta := \epsilon(u_\delta)$ denotes the strain of the regularized system. Our further analysis makes also use of the truncated strain $\widehat{e}_\delta$ (the strain in the not completely damaged parts of Ω_T) given by

$$\widehat{e}_\delta := e_\delta \mathbb{1}_{\{z_\delta > 0\}}.$$

We proceed by deriving suitable a-priori estimates for the incomplete damage problem with respect to δ.

Lemma 6.3.10 (A-priori estimates) *There exists a $C > 0$ independent of δ such that*

(i) $\|\widehat{e}_\delta\|_{L^2(\Omega_T;\mathbb{R}^{n\times n})} \leq C$,   *(iii) $\|\partial_t z_\delta\|_{L^2(\Omega_T)} \leq C$,*

(ii) $\sup_{t\in[0,T]} \|z_\delta(t)\|_{W^{1,p}(\Omega)} \leq C$,   *(iv) $\|W_\delta^{\mathrm{el}}(e_\delta, z_\delta)\|_{L^\infty(0,T;L^1(\Omega))} \leq C$.*

Proof. Applying Gronwall's lemma to the energy estimate (6.20) and noticing the boundedness of $\mathcal{E}_\delta(\epsilon(u_\delta^0), z_\delta^0)$ with respect to $\delta \in (0,1)$ show (iii) and

$$\mathcal{E}_\delta(e_\delta(t), z_\delta(t)) \leq C \tag{6.22}$$

for a.e. $t \in (0,T)$ and all $\delta \in (0,1)$ (cf. Chapter 4) as well as the claim (iv). Taking the restriction $0 \leq z_\delta \leq 1$ into account, property (6.22) gives rise to $\|z_\delta\|_{L^\infty(0,T;W^{1,p}(\Omega))} \leq C$. Together with the control of the time-derivative (iii), we obtain boundedness of $\|z_\delta(t)\|_{W^{1,p}(\Omega)} \leq C$ for every $t \in [0,T]$ and $\delta \in (0,1)$. Hence, (ii) is proven.

It remains to show (i). To proceed, we test inequality (6.19) with $\zeta \equiv -1$ and integrate from $t = 0$ to $t = T$:

$$\int_{\Omega_T} \left(\partial_z W_\delta^{el}(e_\delta, z_\delta) + r_\delta \right) dx\, dt \leq \int_{\Omega_T} (\alpha - \beta\, \partial_t z_\delta)\, dx\, dt. \tag{6.23}$$

Applying (6.2), (6.21) and (6.23), yield

$$
\begin{aligned}
\int_{\Omega_T} \eta |\widehat{e}_\delta|^2\, dx\, dt &= \int_{\{z_\delta > 0\}} \eta |e_\delta|^2\, dx\, dt \\
&\leq \int_{\{z_\delta > 0\}} \frac{1}{2} g'(z_\delta) \mathbf{C} e_\delta : e_\delta\, dx\, dt \\
&= \int_{\Omega_T} \partial_z W_\delta^{el}(e_\delta, z_\delta)\, dx\, dt - \int_{\{z_\delta = 0\}} \partial_z W_\delta^{el}(e_\delta, z_\delta)\, dx\, dt \\
&\leq \int_{\Omega_T} \partial_z W_\delta^{el}(e_\delta, z_\delta)\, dx\, dt - \int_{\Omega_T} \chi_\delta \partial_z W_\delta^{el}(e_\delta, z_\delta)\, dx\, dt \\
&= \int_{\Omega_T} \left(\partial_z W_\delta^{el}(e_\delta, z_\delta) + r_\delta(t) \right) dx\, dt \\
&\leq \int_{\Omega_T} (\alpha - \beta\, \partial_t z_\delta)\, dx\, dt.
\end{aligned}
$$

This and the boundedness of $\int_{\Omega_T} (\alpha - \beta\, \partial_t z_\delta)\, dx\, dt$ with respect to δ show (i). $\qquad\square$

Lemma 6.3.11 (Converging subsequences) *There exists functions*

$$\widehat{e} \in L^2(\Omega_T; \mathbb{R}^{n \times n}), \quad z \in L^\infty(0,T;W^{1,p}(\Omega)) \cap H^1(0,T;L^2(\Omega)),$$

where z is monotonically decreasing with respect to t, i.e., $\partial_t z \leq 0$, and a subsequence (we omit the index) such that for $\delta \to 0^+$:

(i) $z_\delta \rightharpoonup z$ *in* $H^1(0,T;L^2(\Omega))$,   *(ii)* $\widehat{e}_\delta \rightharpoonup \widehat{e}$ *in* $L^2(\Omega_T; \mathbb{R}^{n \times n})$,

    $z_\delta \to z$ *in* $L^p(0,T;W^{1,p}(\Omega))$, $\partial_e W_\delta^{el}(e_\delta, z_\delta) \to \partial_e W^{el}(\widehat{e}, z)$ *in* $L^2(\{z > 0\}; \mathbb{R}^{n \times n})$,

    $z_\delta(t) \to z(t)$ *in* $W^{1,p}(\Omega)$, $\partial_e W_\delta^{el}(e_\delta, z_\delta) \to 0$ *in* $L^2(\{z = 0\}; \mathbb{R}^{n \times n})$.

    $z_\delta \to z$ *in* $\overline{\Omega_T}$,

Proof. The a-priori estimates from Lemma 6.3.10 and classical compactness theorems as well as compactness theorems from Lions and Aubin yield (see Theorem 2.3.9 (i))

$$z_\delta \overset{*}{\rightharpoonup} z \text{ in } L^\infty(0,T;W^{1,p}(\Omega)), \qquad \widehat{e}_\delta \rightharpoonup \widehat{e} \text{ in } L^2(\Omega_T;\mathbb{R}^{n\times n}),$$
$$z_\delta \rightharpoonup z \text{ in } H^1(0,T;L^2(\Omega)), \qquad \partial_e W_\delta^{\mathrm{el}}(e_\delta, z_\delta) \rightharpoonup w_e \text{ in } L^2(\Omega_T;\mathbb{R}^{n\times n}),$$
$$z_\delta \to z \text{ in } L^p(\Omega_T)$$

as $\delta \to 0^+$ for a subsequence and appropriate functions w_e, $\widehat{e}$ and z.

Proving the strong convergence of ∇z_δ in $L^p(\Omega_T;\mathbb{R}^n)$ does not substantially differ from the proof presented in Lemma 4.3.8. It is essentially based on the elementary inequality

$$C_{\mathrm{uc}}|x-y|^p \le \left\langle (|x|^{p-2}x - |y|^{p-2}y), x-y \right\rangle,$$

where $\langle \cdot, \cdot \rangle$ denotes the standard Euclidean scalar product and on an approximation scheme $\{\zeta_\delta\} \subseteq L^p(0,T;W^{1,p}(\Omega))$ with $\zeta_\delta \ge 0$ and

$$\zeta_\delta \to z \text{ in } L^p(0,T;W^{1,p}(\Omega)) \text{ as } \delta \to 0^+, \tag{6.24a}$$
$$0 \le \zeta_\delta \le z_\delta \text{ a.e. in } \Omega_T \text{ for all } \delta \in (0,1). \tag{6.24b}$$

Using the above properties, we obtain the estimate:

$$
C_{\mathrm{uc}} \int_{\Omega_T} |\nabla z_\delta - \nabla z|^p \, \mathrm{d}x \, \mathrm{d}t
$$
$$
\le \int_{\Omega_T} (|\nabla z_\delta|^{p-2}\nabla z_\delta - |\nabla z|^{p-2}\nabla z) \cdot \nabla(z_\delta - z) \, \mathrm{d}x \, \mathrm{d}t
$$
$$
= \underbrace{\int_{\Omega_T} |\nabla z_\delta|^{p-2}\nabla z_\delta \cdot \nabla(z_\delta - \zeta_\delta) \, \mathrm{d}x \, \mathrm{d}t}_{A_\delta}
$$
$$
+ \underbrace{\int_{\Omega_T} \left(|\nabla z_\delta|^{p-2}\nabla z_\delta \cdot \nabla(\zeta_\delta - z) - |\nabla z|^{p-2}\nabla z \cdot \nabla(z_\delta - z)\right) \mathrm{d}x \, \mathrm{d}t}_{B_\delta}.
$$

The weak convergence property of $\{\nabla z_\delta\}$ in $L^p(\Omega_T;\mathbb{R}^n)$ and (6.24a) show $B_\delta \to 0$ as $\delta \to 0^+$. Property (6.19) tested with $\zeta(t) = \zeta_\delta(t) - z_\delta(t)$ and integration from $t = 0$ to $t = T$ yields

$$
A_\delta \le \underbrace{\int_{\Omega_T} \partial_z W_\delta^{\mathrm{el}}(\epsilon(u_\delta), z_\delta)(\zeta_\delta - z_\delta) \, \mathrm{d}x \, \mathrm{d}t}_{\le 0 \text{ by (6.2) and (6.24b)}} + \underbrace{\int_{\Omega_T} (-\alpha + \beta(\partial_t z_\delta(t)))(\zeta_\delta - z_\delta) \, \mathrm{d}x \, \mathrm{d}t}_{\to 0 \text{ as } \delta \to 0^+ \text{ by (6.24a)}}.
$$

Here, we have used $r_\delta\zeta = 0$ in Ω_T (see (6.21)). Therefore, (i) is also shown.

To prove (ii), we define N_δ to be $\{z_\delta > 0\} \cap \{z > 0\}$. Consequently, we get

$$\partial_e W_\delta^{\mathrm{el}}(\widehat{e}_\delta, z_\delta)\mathbb{1}_{N_\delta} = \partial_e W_\delta^{\mathrm{el}}(e_\delta, z_\delta)\mathbb{1}_{N_\delta} \tag{6.25}$$

and the convergence

$$\mathbb{1}_{N_\delta} \to \mathbb{1}_{\{z>0\}} \text{ in } \Omega_T \qquad (6.26)$$

for $\delta \to 0^+$ by using $z_\delta \to z$ in $\overline{\Omega_T}$. Calculating the weak $L^1(\Omega_T; \mathbb{R}^{n\times n})$-limits in (6.25) for $\delta \to 0^+$ on both sides by using the already proven convergence properties, we obtain $\partial_e W^{\mathrm{el}}(\hat{e}, z) = w_e$. The remaining convergence property in (ii) follows from Lemma 6.3.10 (iv). $\qquad \square$

We now introduce the shrinking set $F \subseteq \overline{\Omega_T}$ by defining

$$F(t) := \{z(t) > 0\}$$

for all $t \in [0, T]$. This is a well-defined object since $F \subseteq \overline{\Omega_T}$ is relatively open by Proposition 2.3.14 as well as $F(s) \subseteq F(t)$ for all $0 \le t \le s \le T$ by the monotone decrease of $z(x, \cdot)$.

Corollary 6.3.12 *Let $t \in [0, T]$ and $U \subset\subset F(t)$ be an open subset. Then $U \subseteq \{z_\delta(s) > 0\}$ for all $s \in [0, t]$ provided that $\delta > 0$ is sufficiently small. More precisely, there exist $0 < \delta_0, \eta < 1$ such that*

$$z_\delta(s) \ge \eta \text{ in } U$$

for all $s \in [0, t]$ and for all $0 < \delta < \delta_0$.

Proof. By assumption, we obtain the property $\mathrm{dist}(U, \{z(t) = 0\}) > 0$. Therefore, and by $z(t) \in \mathcal{C}(\overline{\Omega})$, we find an $\eta > 0$ such that $z(t) \ge 2\eta$ in U. By exploiting the convergence $z_\delta(t) \to z(t)$ in $\mathcal{C}(\overline{\Omega})$ as $\delta \to 0^+$ by Lemma 6.3.11 (b) and the compact embedding $W^{1,p}(\Omega) \hookrightarrow \mathcal{C}(\overline{\Omega})$, there exists an $\delta_0 > 0$ such that $z_\delta(t) \ge \eta$ on U for all $0 < \delta < \delta_0$. Finally, the claim follows from the fact that z_δ is monotonically decreasing with respect to t. $\qquad \square$

Lemma 6.3.13 *There exists a function $u \in L_t^2 H_{x,\mathrm{loc}}^1(\mathfrak{A}_{\Gamma_\mathrm{D}}(F); \mathbb{R}^n)$ such that*

(i) $\epsilon(u) = \hat{e}$ a.e. in $\mathfrak{A}_{\Gamma_\mathrm{D}}(F)$,

(ii) $u = b$ on the boundary $(\Gamma_\mathrm{D})_T \cap \mathfrak{A}_{\Gamma_\mathrm{D}}(F)$.

Proof. Let $\{U_k^m\}$ and $\{t_m\}$ be sequences satisfying the properties of Corollary 2.4.7 applied to $\mathfrak{A}_{\Gamma_\mathrm{D}}(F)$. We get for each fixed $k, m \in \mathbb{N}$

$$U_k^m \times [0, t_m] \subseteq \{z_\delta > 0\} \qquad (6.27)$$

for all $0 < \delta \ll 1$ due to Corollary 6.3.12. Inclusion (6.27) implies

$$\epsilon(u_\delta) = \hat{e}_\delta \qquad (6.28)$$

a.e. in $U_k^m \times (0, t_m)$. Korn's inequality applied on the Lipschitz domain U_k^m yields (note that $\mathcal{H}^{n-1}(\partial U_k^m \cap \Gamma_D) > 0$)

$$\|u_\delta\|_{L^2(0,t_m;H^1(U_k^m;\mathbb{R}^n))}^2 \leq 2 \int_0^{t_m} \|u_\delta(t) - b(t)\|_{H^1(U_k^m;\mathbb{R}^n)}^2 + \|b(t)\|_{H^1(U_k^m;\mathbb{R}^n)}^2 \, dt$$

$$\leq C \left(1 + \int_0^{t_m} \|\epsilon(u_\delta(t))\|_{L^2(U_k^m;\mathbb{R}^{n\times n})}^2 \, dt \right)$$

$$\leq C \left(1 + \int_0^{t_m} \|\widehat{e}_\delta(t)\|_{L^2(\Omega;\mathbb{R}^{n\times n})}^2 \, dt \right)$$

with a constant $C = C(U_k^m, b) > 0$. Together with the boundedness of $\widehat{e}_\delta$ in $L^2(\Omega_T; \mathbb{R}^{n\times n})$, we can find a subsequence $\delta \to 0^+$ and a function $u^{(k,m)} \in L^2(0, t_m; H^1(U_k^m; \mathbb{R}^n))$ such that

$$u_\delta \rightharpoonup u^{(k,m)} \text{ in } L^2(0, t_m; H^1(U_k^m; \mathbb{R}^n)). \tag{6.29}$$

Thus $\epsilon(u^{(k,m)}) = \widehat{e}$ in $U_k^m \times (0, t_m)$ because of (6.28) and the weak convergence property of $\widehat{e}_\delta$. For each $k, m \in \mathbb{N}$, we can apply the argumentation above. Therefore, by successively choosing subsequences and by applying a diagonalization argument, we obtain a subsequence $\delta \to 0^+$ such that (6.29) holds for all $k, m \in \mathbb{N}$.

Since $u^{(k_1,m_1)} = u^{(k_2,m_2)}$ a.e. on $U_{k_1}^{m_1} \times (0, t_{m_1}) \cap U_{k_2}^{m_2} \times (0, t_{m_2})$ for all $k_1, k_2, m_1, m_2 \in \mathbb{N}$, we obtain an $u : \mathfrak{A}_{\Gamma_D}(F) \to \mathbb{R}^n$ such that $u|_{U_k^m \times (0,t_m)} \in L^2(0, t_m; H^1(U_k^m; \mathbb{R}^n))$ for all $m \in \mathbb{N}$. Proposition 2.4.10 (a) yields $u \in L_t^2 H_{x,\text{loc}}^1(F; \mathbb{R}^n)$ and the symmetric gradient $\epsilon(u)$ coincides with $\widehat{e}$. Therefore, (i) is shown.

Furthermore, for every $k, m \in \mathbb{N}$, we have $u(t) = b(t)$ on $\partial U_k^m \cap \Gamma_D$ in the sense of traces for a.e. $t \in [0, t_m]$. By Proposition 2.4.10 (b), (ii) follows. $\qquad\square$

We are now able to prove Proposition 6.3.7.

Proof of Proposition 6.3.7. Lemma 6.3.11 and Lemma 6.3.13 give the desired regularity properties of the functions (e, u, z) in Proposition 6.3.7. Here, we set $e := \widehat{e}|_F \in L^2(F; \mathbb{R}^{n\times n})$. The property $e = \epsilon(u)$ in $\mathfrak{A}_{\Gamma_D}(F)$ follows from Lemma 6.3.13.

In the following, we are going to prove that properties (ii)-(v) of Definition 6.2.1 are satisfied.

(ii) Lemma 6.3.11 (ii) allows us to pass to $\delta \to 0^+$ in (6.18) integrated from $t = 0$ to $t = T$. Therefore, equation (6.4) holds for a.e. $t \in (0, T)$ and all $\zeta \in H_{\Gamma_D}^1(\Omega; \mathbb{R}^n)$. Moreover, the boundary condition $u = b$ on $(\Gamma_D)_T \cap \mathfrak{A}_{\Gamma_D}(F)$ is satisfied. Definition 2.4.3 immediately implies $(\Gamma_D)_T \cap F = (\Gamma_D)_T \cap \mathfrak{A}_{\Gamma_D}(F)$.

(iii) We first show (6.5). Let $\zeta \in L^\infty(0, T; W^{1,p}(\Omega))$ with $\zeta \leq 0$. The variational inequality (6.19) and the representation for r_δ (6.21) imply

$$0 \leq \int_{\Omega_T} \left(|\nabla z_\delta|^{p-2} \nabla z_\delta \cdot \nabla \zeta + (-\alpha + \beta \partial_t z_\delta) \zeta \right) dx \, dt + \int_{\{z_\delta > 0\}} \partial_z W_\delta^{\text{el}}(e_\delta, z_\delta) \zeta \, dx \, dt. \tag{6.30}$$

In addition,

$$\int_{\{z_\delta>0\}} \partial_z W_\delta^{\mathrm{el}}(e_\delta, z_\delta)\zeta \,\mathrm{d}x\,\mathrm{d}t \leq \int_{F\cap\{z_\delta>0\}} \partial_z W_\delta^{\mathrm{el}}(e_\delta, z_\delta)\zeta \,\mathrm{d}x\,\mathrm{d}t$$

$$= \int_F g'(z_\delta)\mathbf{C}\widehat{e}_\delta : \widehat{e}_\delta\zeta \,\mathrm{d}x\,\mathrm{d}t.$$

Lemma 6.3.11, a lower semi-continuity argument and $\mathbb{1}_{\{z_\delta>0\}\cap\{z=0\}} \to \mathbb{1}_{\{z=0\}}$ a.e. in Ω_T (see proof of Lemma 6.3.11) yield

$$\limsup_{\delta\to0^+} \int_{\{z_\delta>0\}} \partial_z W_\delta^{\mathrm{el}}(e_\delta, z_\delta)\zeta \,\mathrm{d}x\,\mathrm{d}t \leq \int_F \partial_z W_\delta^{\mathrm{el}}(e, z)\zeta \,\mathrm{d}x\,\mathrm{d}t.$$

Therefore, applying "$\limsup_{\delta\to0^+}$" on both sides of (6.30), using the above estimate and Lemma 6.3.11 yield

$$\int_F \left(|\nabla z|^{p-2}\nabla z \cdot \nabla\zeta + \partial_z W^{\mathrm{el}}(e, z)\zeta\right) \mathrm{d}x \geq \int_\Omega (\alpha - \beta\partial_t z)\zeta \,\mathrm{d}x. \tag{6.31}$$

The properties $\partial_t z \leq 0$ and $z \geq 0$ a.e. in Ω_T follow from Lemma 6.3.11 by taking $\partial_t z_\delta \leq 0$ and $z_\delta \geq 0$ a.e. in Ω_T into account.

(iv) The jump condition (6.6) in (iv) of Definition 6.2.1 holds trivially since we have the regularity $z \in L^\infty(0, T; W^{1,p}(\Omega)) \cap H^1(0, T; L^2(\Omega))$.

(v) To complete the proof, we need to show the energy estimates (6.7). Since $\{b^0, z_\delta^0\}$ is a recovery sequence, we get $\mathcal{E}_\delta(\epsilon(u_\delta^0), z_\delta^0) \to \mathfrak{E}(b^0, z^0)$ as $\delta \to 0^+$. Now, applying "$\limsup_{\delta\to0^+}$" on both sides in (6.20) and using the convergence properties in Lemma 6.3.11 as well as lower semi-continuity arguments yield

$$\mathfrak{E}(b^0, z^0) + \int_0^t \int_{F(s)} \partial_e W^{\mathrm{el}}(e, z) : \epsilon(\partial_t b) \,\mathrm{d}x\,\mathrm{d}s$$

$$\geq \limsup_{\delta\to0^+} \left(\mathcal{E}_\delta(e_\delta(t), z_\delta(t)) + \int_{\Omega_t} \left(\alpha|\partial_t z_\delta| + \beta|\partial_t z_\delta|^2\right) \mathrm{d}x\,\mathrm{d}s\right)$$

$$\geq \limsup_{\delta\to0^+} \int_\Omega W_\delta^{\mathrm{el}}(e_\delta(t), z_\delta(t)) \,\mathrm{d}x + \int_\Omega \frac{1}{p}|\nabla z(t)|^p \,\mathrm{d}x$$

$$+ \int_{\Omega_t} \left(\alpha|\partial_t z| + \beta|\partial_t z|^2\right) \mathrm{d}x\,\mathrm{d}s. \tag{6.32}$$

Indeed, for an arbitrary $t \in (0, T)$, we derive by Fatou's lemma and Lemma 6.3.11

$$\int_0^t \left(\limsup_{\delta\to0^+} \int_\Omega W_\delta^{\mathrm{el}}(e_\delta(s), z_\delta(s)) \,\mathrm{d}x\right) \mathrm{d}s \geq \limsup_{\delta\to0^+} \int_{\Omega_t} W_\delta^{\mathrm{el}}(e_\delta, z_\delta) \,\mathrm{d}x\,\mathrm{d}s$$

$$\geq \liminf_{\delta\to0^+} \int_F (g(z_\delta) + \delta)\mathbf{C}\widehat{e}_\delta : \widehat{e}_\delta \,\mathrm{d}x\,\mathrm{d}s$$

$$\geq \int_F W^{\mathrm{el}}(e, z) \,\mathrm{d}x\,\mathrm{d}s. \tag{6.33}$$

Here, we have used the weak convergence property

$$\sqrt{g(z_\delta) + \delta}\, \widehat{e}_\delta \rightharpoonup \sqrt{g(z)}\, \widehat{e} \text{ in } L^2(\Omega_T; \mathbb{R}^{n \times n})$$

as $\delta \to 0^+$. To finish the proof, (6.33) implies

$$\limsup_{\delta \to 0^+} \int_\Omega W_\delta^{\text{el}}(e_\delta(t), z_\delta(t))\, \mathrm{d}x \geq \int_{F(t)} W^{\text{el}}(e(t), z(t))\, \mathrm{d}x$$

for a.e. $t \in (0, T)$. Combining it with (6.32), estimate (6.7) is shown. □

6.3.3 Local- and global-in-time existence results

By using the achievements in the previous section and Zorn's lemma, we will prove the main results, Theorem 6.2.6 and Theorem 6.2.7.

The general idea behind the global existence proof is illustrate in Figure 6.2. Starting from an initial damage profile z^0 at time $t_0 = 0$, we calculate a degenerate limit solution via Proposition 6.3.7. Suppose that the first material exclusions occur at time t_1. We define a new initial condition z^1 where the excluded fragments in $z(t_1)$ are set to 0 and we calculate a degenerate limit again. By repeating this procedure, we obtain $t_0 \leq t_1 \leq t_2 \leq \ldots$ The degenerate limit functions on (t_i, t_{i+1}) can be concatenated to a weak solution on $(t_0, \sup_{i \in \mathbb{N}} t_i)$ with a jump term in the energy inequality at each time t_i. However, a problem arises when infinitely many material exclusions occur in an arbitrary short time interval (see Figure 6.1). In this case, we will "neglect small" material

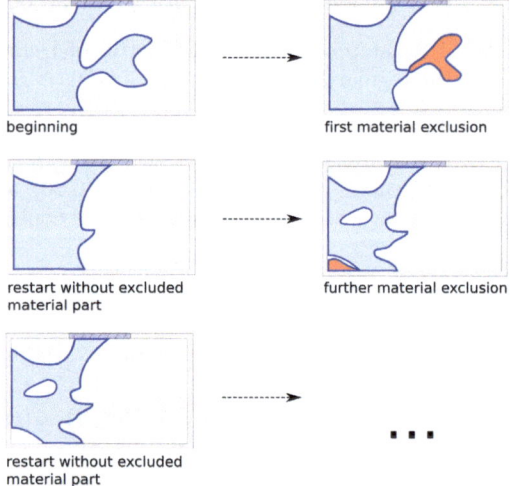

Figure 6.2: *Concatenation of solutions from the degenerate limit.*

exclusions such that we can ensure $t_i < t_{i+1}$ for every $i \in \mathbb{N}$. This leads to the concept of approximate weak solutions introduced in Definition 6.2.4. A further problem might be that $\sup_{i \in \mathbb{N}} t_i < T$ holds. This problem is circumvented by working with arbitrary chains of solutions and using Zorn's lemma instead of complete induction. Maximal elements will then correspond to approximate weak solutions on $(0, T)$.

To proceed, let $\eta > 0$ be fixed and $\mathcal{P}$ be the set

$$\mathcal{P} := \{ (\widehat{T}, e, u, z, F) \,|\, 0 < \widehat{T} \leq T \text{ and } (e, u, z, F) \text{ is an approximate weak solution on}$$
$$[0, \widehat{T}] \text{ with fineness } \eta \text{ according to Definition 6.2.4 } \}.$$

We introduce a partial ordering $\leq$ on $\mathcal{P}$ by

$$(\widehat{T}_1, e_1, u_1, z_1, F_1) \leq (\widehat{T}_2, e_2, u_2, z_2, F_2) \quad \Leftrightarrow \quad \widehat{T}_1 \leq \widehat{T}_2, \ e_2|_{[0,\widehat{T}_1]} = e_1, \ u_2|_{[0,\widehat{T}_1]} = u_1,$$
$$z_2|_{[0,\widehat{T}_1]} = z_1, \ F_2|_{[0,\widehat{T}_1]} = F_1.$$

The next two lemma prove the assumptions for Zorn's lemma.

Lemma 6.3.14 $\mathcal{P} \neq \emptyset$.

Proof. Let (e, u, z) be the tuple from Proposition 6.3.7 to the initial-boundary data (z^0, b). If there exists an $\delta > 0$ such that $J_{z^\star} \cap [0, \delta] = \emptyset$ with $z^\star(t) := z(t) \mathbb{1}_{\mathfrak{A}_{\Gamma_D}(\{z^-(t)>0\})}$ then $(\delta, e, u, z, F) \in \mathcal{P}$. Otherwise, we find $0 \in C_{z^\star}$. We claim

$$\mathcal{L}^n \left(\{z^0 > 0\} \setminus \mathfrak{A}_{\Gamma_D}(\{z(t) > 0\}) \right) \to 0 \text{ as } t \to 0^+. \tag{6.34}$$

We consider the non-trivial case $z^0 \not\equiv 0$. Let $x \in \{z^0 > 0\} \cap \Omega$. Since $\{z^0 > 0\} \subseteq \overline{\Omega_T}$ is relatively open and admissible with respect to Γ_D, there exists a Lipschitz domain $U \subset\subset \{z^0 > 0\}$ with $x \in U$ such that $\mathcal{H}^{n-1}(\partial U \cap \Gamma_D) > 0$ by Lemma 2.4.6. Because of Proposition 2.3.14, $z \in \mathcal{C}(\overline{\Omega_T})$ and, consequently, there exists a $t > 0$ such that $U \subset\subset \{z(s) > 0\}$ for all $0 \leq s < t$. In particular, $x \in \mathfrak{A}_{\Gamma_D}(\{z(s) > 0\})$ for all $0 \leq s < t$. This proves (6.34). Finally, choose $\delta > 0$ so small such that $\delta < \eta$ and (note the monotonicity of z with respect to t)

$$\mathcal{L}^n \left(\{z(t) > 0\} \setminus \mathfrak{A}_{\Gamma_D}(\{z(t) > 0\}) \right) \leq \mathcal{L}^n \left(\{z^0 > 0\} \setminus \mathfrak{A}_{\Gamma_D}(\{z(t) > 0\}) \right) < \eta$$

for all $0 \leq t < \delta$. We have proved that (e, u, z) on $F := \{z > 0\}$ is an approximate weak solution with fineness η on the time interval $[0, \delta]$, i.e., $(\delta, e, u, z, F) \in \mathcal{P}$. $\qquad \square$

Lemma 6.3.15 *Every totally ordered subset of $\mathcal{P}$ has an upper bound.*

Proof. Let $\mathcal{R} \subseteq \mathcal{P}$ be a totally ordered subset. We denote with $[0, T_R]$ the corresponding time interval of an element $R \in \mathcal{R}$. Let us select a sequence $\{T_\theta, e_\theta, u_\theta, z_\theta, F_\theta\}_{\theta \in (0,1)} \subseteq \mathcal{R}$, with $T_{\theta_1} \leq T_{\theta_2}$ for $\theta_2 \leq \theta_1$ and $\lim_{\theta \to 0^+} T_\theta = \sup_{Q \in \mathcal{R}} T_Q =: \widehat{T}$.

Let $t \in (0, \widehat{T})$. There exists a $\theta \in (0, 1)$ with $T_\theta \geq t$ and we define

$$(e(t), u(t), z(t), F(t)) := (e_\theta(t), u_\theta(t), z_\theta(t), F_\theta(t)).$$

By construction, the functions (e, u, z) satisfy the properties (ii)-(v) of Definition 6.2.1 on $[0, \widehat{T}]$. It remains to show that $(e(t), u(t), z(t))$ are in the trajectory spaces required as in Definition 6.2.4 (i) and that F satisfies Definition 6.2.4 (ii).

The energy estimate for $(e_\theta, u_\theta, z_\theta)$ implies

$$\mathcal{E}(e(t), z(t)) + \int_0^t \int_{F(s)} \left(\alpha |\partial_t^a z| + \beta |\partial_t^a z|^2 \right) \mathrm{d}x \, \mathrm{d}s$$

$$\leq \mathfrak{e}_0^+ + \int_0^t \int_{F(s)} \partial_e W^{\mathrm{el}}(e, z) : \epsilon(\partial_t b) \, \mathrm{d}x \, \mathrm{d}s \qquad (6.35)$$

for a.e. $t \in (0, \widehat{T})$. Gronwall's lemma yields boundedness of the left hand side of (6.35) with respect to a.e. $t \in (0, \widehat{T})$.

We immediately get

$$z \in L^\infty(0, \widehat{T}; W^{1,p}(\Omega)) \cap SBV^2(0, \widehat{T}; L^2(\Omega)). \qquad (6.36)$$

Variational inequality (6.5) tested with $\zeta \equiv -1$ shows

$$\int_{F(t)} \partial_z W^{\mathrm{el}}(e(t), z(t)) \, \mathrm{d}x \leq \int_\Omega \alpha \, \mathrm{d}x - \int_{F(t)} \beta \partial_t^a z(t) \, \mathrm{d}x$$

for a.e. $t \in (0, \widehat{T})$. This implies

$$e \in L^2(F; \mathbb{R}^{n \times n}). \qquad (6.37)$$

We know that $u|_{U \times (0,t)} \in L^2(0, t; H^1(U; \mathbb{R}^n))$ for all $t \in (0, \widehat{T})$ and all open subsets $U \subset\subset \mathfrak{A}_{\Gamma_{\mathrm{D}}}(F(t))$. Let $\{U_k\}$ be a Lipschitz cover of the admissible set

$$F(\widehat{T}) := \mathfrak{A}_{\Gamma_{\mathrm{D}}}(\{z^-(\widehat{T}) > 0\})$$

according to Lemma 2.4.6 (in particular, Definition 6.2.4 (ii) is fulfilled). For each $k \in \mathbb{N}$, we apply Korn's inequality and get for all $t \in (0, \widehat{T})$

$$\|u - b\|_{L^2(0,t;H^1(U_k;\mathbb{R}^n))} \leq C \|\epsilon(u)\|_{L^2(0,t;L^2(U_k;\mathbb{R}^n))},$$

where $C > 0$ depends on the domain U_k but not on the time t. Thus $u|_{U_k \times (0,\widehat{T})} \in L^2(0, T; H^1(U_k; \mathbb{R}^n))$. In conclusion,

$$u \in L_t^2 H_{x,\mathrm{loc}}^1(F; \mathbb{R}^n). \qquad (6.38)$$

Therefore, property (i) of Definition 6.2.4 follows by (6.36)-(6.38). We end up with $\{\widehat{T}, e, u, z, F\} \in \mathcal{P}$ satisfying $\{T_\theta, e_\theta, u_\theta, z_\theta, F_\theta\} \leq \{\widehat{T}, e, u, z, F\}$ for all $\theta \in (0, 1)$. $\qquad \square$

Weak solutions exhibit the following concatenation property.

Lemma 6.3.16 *Let $t_1 < t_2 < t_3$ be real numbers. Suppose that*

$$\widetilde{q} := (\widetilde{e}, \widetilde{u}, \widetilde{z}, \widetilde{F}) \text{ is an approximate weak solution on } [t_1, t_2],$$

$\widehat{q} := (\widehat{e}, \widehat{u}, \widehat{z}, \widehat{F})$ *is an approximate weak solution on* $[t_2, t_3]$

with $\widehat{\mathfrak{e}}_{t_2}^+ = \mathfrak{E}(\widehat{b}(t_2), \widehat{z}^+(t_2))$ *(the value* $\mathfrak{e}_{t_2}^+$ *for* $\widehat{q}$ *in Definition 6.2.1).*

Furthermore, suppose the compatibility condition $\widehat{z}^+(t_2) = \widetilde{z}^-(t_2)\mathbb{1}_{\mathfrak{A}_{\Gamma_D}(\{\widetilde{z}^-(t_2)>0\})}$ *and the Dirichlet boundary data* $b \in W^{1,1}(t_1, t_3; W^{1,\infty}(\Omega; \mathbb{R}^n))$. *Then, we obtain that* $q := (e, u, z, F)$ *defined as* $q|_{[t_1, t_2)} := \widetilde{q}$ *and* $q|_{[t_2, t_3]} := \widehat{q}$ *is an approximate weak solution on* $[t_1, t_3]$.

Proof. Applying "$\lim_{s \to t_2^-} \operatorname{ess\,inf}_{\tau \in (s, t_2)}$" on both sides of the energy estimate (6.7) for $(\widetilde{e}, \widetilde{u}, \widetilde{z}, \widetilde{F})$ yields

$$\lim_{s \to t_2^-} \operatorname{ess\,inf}_{\tau \in (s, t_2)} \mathcal{E}(e(\tau), z(\tau)) + \int_{t_1}^{t_2} \int_{F(s)} (\alpha |\partial_t^a z| + \beta |\partial_t^a z|^2)\, dx\, ds + \liminf_{s \to t_2^-} \sum_{\tau \in J_z \cap (t_1, s]} \mathcal{J}_\tau$$

$$\leq \mathfrak{e}_{t_1}^+ + \int_{t_1}^{t_2} \int_{F(s)} \partial_e W^{el}(e, z) : \epsilon(\partial_t b)\, dx\, ds.$$

This estimate can be rewritten as

$$\mathfrak{E}(b(t_2), z^+(t_2)) + \int_{t_1}^{t_2} \int_{F(s)} (\alpha |\partial_t^a z| + \beta |\partial_t^a z|^2)\, dx\, ds$$

$$+ \liminf_{s \to t_2^-} \sum_{\tau \in J_z \cap (t_2, s]} \mathcal{J}_\tau + \lim_{s \to t_2^-} \operatorname{ess\,inf}_{\tau \in (s, t_2)} \mathcal{E}(e(\tau), z(\tau)) - \mathfrak{E}(b(t_2), z^+(t_2))$$

$$\leq \mathfrak{e}_{t_1}^+ + \int_{t_1}^{t_2} \int_{F(s)} \partial_e W^{el}(e, z) : \epsilon(\partial_t b)\, dx\, ds. \tag{6.39}$$

In the following, we show that we may choose the value $\mathfrak{E}(b(t_2), z^+(t_2))$ for $\mathfrak{e}_{t_2}^+$. By the property (i) of Definition 6.2.1, we get $z^-(s) \rightharpoonup z^-(t_2)$ in $W^{1,p}(\Omega)$ and $b(s) \to b(t_2)$ in $W^{1,\infty}(\Omega; \mathbb{R}^n)$ as $s \to t_2^-$. In particular, by using Lemma 6.3.5 and the monotone decrease of z^- with respect to t,

$$z^-(s)\mathbb{1}_{\mathfrak{A}_{\Gamma_D}(\{z^-(s)>0\})} \rightharpoonup z^-(t_2)\mathbb{1}_{\bigcap_{\tau \in (t_1, t_2)} \mathfrak{A}_{\Gamma_D}(\{z^-(\tau)>0\})} =: \chi$$

in $W^{1,p}(\Omega)$ as $s \to t_2^-$. By the definition of χ, the inclusion

$$\mathfrak{A}_{\Gamma_D}(\{z^-(t_2) > 0\}) \subseteq \bigcap_{\tau \in (t_1, t_2)} \mathfrak{A}_{\Gamma_D}(\{z^-(\tau) > 0\})$$

and the compatibility condition, we find $z^+(t_2) = \chi \mathbb{1}_{\mathfrak{A}_{\Gamma_D}(\{z^-(t_2)>0\})}$.

Thus, applying Lemma 6.3.6, lower semi-continuity of the Γ-limit $\mathfrak{E}$ and Corollary 6.3.4 (iii), we obtain

$$\lim_{s \to t_2^-} \operatorname{ess\,inf}_{\tau \in (s, t_2)} \mathcal{E}(e(\tau), z(\tau)) = \lim_{s \to t_2^-} \operatorname{ess\,inf}_{\tau \in (s, t_2)} \mathcal{E}(e(\tau), z^-(\tau))$$

$$\geq \lim_{s \to t_2^-} \operatorname*{ess\,inf}_{\tau \in (s,t_2)} \mathcal{E}(\epsilon(u(\tau)), z^-(\tau) \mathbb{1}_{\mathfrak{A}_{\Gamma_D}}(\{z^-(\tau)>0\}))$$

$$\geq \lim_{s \to t_2^-} \operatorname*{ess\,inf}_{\tau \in (s,t_2)} \mathfrak{E}(b(\tau), z^-(\tau) \mathbb{1}_{\mathfrak{A}_{\Gamma_D}}(\{z^-(\tau)>0\}))$$

$$\geq \mathfrak{E}(b(t_2), \chi)$$

$$\geq \mathfrak{E}(b(t_2), z^+(t_2)).$$

This leads to

$$0 \leq \sum_{s \in J_z \cap (t_1,t_2]} \mathcal{J}_s \leq \lim_{s \to t_2^-} \operatorname*{ess\,inf}_{\tau \in (s,t_2)} \mathcal{E}(e(\tau), z(\tau)) - \mathfrak{E}(b(t_2), z^+(t_2))$$

$$+ \liminf_{s \to t_2^-} \sum_{\tau \in J_z \cap (t_1,s]} \mathcal{J}_\tau,$$

where the second '$\leq$' becomes an '$=$' if $t_2 \in J_z$. Consequently, (6.39) becomes

$$\mathfrak{E}(b(t_2), z^+(t_2)) + \int_{t_1}^{t_2} \int_{F(s)} \left(\alpha|\partial_t^a z| + \beta|\partial_t^a z|^2\right) \mathrm{d}x\,\mathrm{d}s + \sum_{s \in J_z \cap (t_1,t_2]} \mathcal{J}_s \tag{6.40}$$

$$\leq \mathfrak{e}_{t_1}^+ + \int_{t_1}^{t_2} \int_{F(s)} \partial_e W^{\mathrm{el}}(e, z) : \epsilon(\partial_t b)\,\mathrm{d}x\,\mathrm{d}s.$$

The energy inequality (6.7) for $(\widehat{e}, \widehat{u}, \widehat{z}, \widehat{F})$ (taking $\widehat{\mathfrak{e}}_{t_2}^+ = \mathfrak{E}(\widehat{b}(t_2), \widehat{z}^+(t_2))$ into account) can be expressed as

$$\mathcal{E}(e(t), z(t)) + \int_{t_2}^{t} \int_{F(s)} \left(\alpha|\partial_t^a z| + \beta|\partial_t^a z|^2\right) \mathrm{d}x\,\mathrm{d}s + \sum_{s \in J_z \cap (t_2,t]} \mathcal{J}_s \tag{6.41}$$

$$\leq \mathfrak{E}(b(t_2), z^+(t_2)) + \int_{t_2}^{t} \int_{F(s)} \partial_e W^{\mathrm{el}}(e, z) : \epsilon(\partial_t b)\,\mathrm{d}x\,\mathrm{d}s$$

for a.e. $t \in (t_2, t_3)$. Adding (6.40) and (6.41) shows that the energy estimate for (e, u, z, F) also holds for a.e. $t \in (t_2, t_3)$. It is now easy to verify that (e, u, z, F) is a approximate weak solution on the time interval $[t_1, t_3]$ according to Definition 6.2.1. $\square$

Proof of Theorem 6.2.6. By Zorn's lemma, we deduce the existence of a maximal element $R = (\widetilde{T}, \widetilde{e}, \widetilde{u}, \widetilde{z}, \widetilde{F})$ in $\mathcal{P}$. In particular, a maximal element satisfies the properties in Theorem 6.2.6 on the interval $[0, \widetilde{T}]$. We deduce $T = \widetilde{T}$. Otherwise, we get another approximate weak solution $(\widehat{e}, \widehat{u}, \widehat{z}, \widehat{F})$ on $[\widetilde{T}, \widetilde{T} + \varepsilon]$ for an $\varepsilon > 0$ with initial datum $\widetilde{z}^-(\widetilde{T}) \mathbb{1}_{\mathfrak{A}_{\Gamma_D}}(\widetilde{z}-(\widetilde{T})>0)$ (which is an element of $W^{1,p}(\Omega)$ by Lemma 6.3.5) as in the proof of Lemma 6.3.14 with $e_{\widetilde{T}}^\pm = \mathfrak{E}(b(\widetilde{T}), z(\widetilde{T}))$ if $\widetilde{T} \in J_z$. By Lemma 6.3.16, $(\widetilde{e}, \widetilde{u}, \widetilde{z}, \widetilde{F})$ and $(\widehat{e}, \widehat{u}, \widehat{z}, \widehat{F})$ can be concatenate to an approximate weak solution on $[0, \widetilde{T} + \varepsilon]$ which is a contradiction. $\square$

Proof of Theorem 6.2.7. Here, let us consider the set $\mathcal{P}$ given by

$$\mathcal{P} := \big\{ (\widehat{T}, u, z) \,\big|\, 0 < \widehat{T} \leq T \text{ and } (u, z) \text{ is a weak solution on}$$
$$[0, \widehat{T}] \text{ according to Definition 6.2.1} \big\}$$

with an ordering $\leq$ as above (except the conditions $e_2|_{[0,\widehat{T}_1]} = e_1$ and $F_2|_{[0,\widehat{T}_1]} = F_1$ which are not needed here). Proposition 6.3.7 shows $\mathcal{P} \neq \emptyset$ by noticing $z \in \mathcal{C}(\overline{\Omega_T})$ (see Proposition 2.3.14) and $0 < \eta \leq z^0$. The property that every totally ordered subset of $\mathcal{P}$ has an upper bound can be shown as in Lemma 6.3.15. A maximal element satisfies the claim. $\square$

Cahn-Hilliard systems coupled with complete damage processes and homogeneous elasticity

This chapter combines the approach from Chapter 4 with the ideas from the preceding chapter. More specifically, we are going to investigate existence of weak solutions for complete damage systems which are coupled with degenerating Cahn-Hilliard equations. The diffusion mobility tensor depends on the damage variable and vanishes when the damage is maximal. Therefore, we have two degenerating terms in the resulting system: the elastic energy density and the mobility tensor.

As in the previous chapter, we will prove local-in-time existence of weak solutions and global-in-time existence of solutions in a weaker sense.

The results and proofs in this chapter can also be found in WIAS preprint no. 1759, see [HK12b].

7.1   Assumptions

Let $\Omega \subseteq \mathbb{R}^n$ be a bounded $\mathcal{C}^2$-domain and $\Gamma_D \subseteq \partial\Omega$ be the Dirichlet boundary with $\mathcal{H}^{n-1}(\Gamma_D) > 0$. Note that in the preceding chapters we have assumed that Ω is a bounded Lipschitz domain. The additional regularity of the boundary is needed in this chapter in order to prove a priori estimates for the degenerating chemical potential μ near the boundary via the conical Poincaré inequality. To keep the presentation short, we assume in this chapter that the alloy is a binary mixture such that c reduces to a scalar function (see Chapter 4).

The free energy density function ψ and the dissipation potential density function ϕ are given in (1.3) and (1.4) with the functions $W^{\text{el}} \in \mathcal{C}^1(\mathbb{R} \times \mathbb{R}^{n \times n} \times \mathbb{R}; \mathbb{R}_+)$, $W^{\text{ch}} \in \mathcal{C}^1(\mathbb{R}; \mathbb{R}_+)$ and $f \in \mathcal{C}^1(\mathbb{R}; \mathbb{R}_+)$, and the exponent p with $p > n$. To keep the presentation short, we assume WLOG $\alpha = 0$ and $\beta = 1$ in (1.4). Moreover, the following product structure for the elastic energy density is supposed:

$$W^{\text{el}}(c, e, z) = g(z)\varphi(c, e), \tag{7.1}$$

where $g \in \mathcal{C}^1([0, 1]; \mathbb{R}_+)$ is a non-negative function which satisfies the conditions

$$\eta \le g'(z), \tag{7.2a}$$
$$g(0) = 0 \tag{7.2b}$$

for all $z \in [0, 1]$ and some constant $\eta > 0$. The incomplete damage case $g(0) > 0$ with constant mobility tensors $\mathbb{M}$ has already been treated in Chapter 5. The second function $\varphi \in \mathcal{C}^1(\mathbb{R} \times \mathbb{R}^{n \times n}_{\text{sym}}; \mathbb{R}_+)$ in (7.1) is assumed to be of the following polynomial form:

$$\varphi(c, e) = \varphi^1 e : e + \varphi^2(c) : e + \varphi^3(c) \tag{7.3}$$

for coefficients $\varphi^1 \in \mathcal{L}(\mathbb{R}^{n \times n}_{\text{sym}})$ with $\varphi^1 > 0$, $\varphi^2 \in \mathcal{C}^1(\mathbb{R}; \mathbb{R}^{n \times n}_{\text{sym}})$ and $\varphi^3 \in \mathcal{C}^1(\mathbb{R})$.

From now on, we suppose that φ and W^{ch} satisfy the growth conditions

$$|\varphi^2(c)|, |\partial_c \varphi^2(c)| \le C(1 + |c|), \tag{7.4a}$$
$$|\varphi^3(c)|, |\partial_c \varphi^3(c)| \le C(1 + |c|^2), \tag{7.4b}$$
$$|\partial_c W^{\text{ch}}(c)| \le C(1 + |c|^{2^*/2}). \tag{7.4c}$$

Here, $C > 0$ denotes a constant independent of c. In the case $n = 2$, $\partial_c W^{\text{ch}}$ has to satisfy an r-growth condition for a fixed arbitrary $r > 0$, whereas we have no restrictions on $\partial_c W^{\text{ch}}$ in the one-dimensional case.

Remark 7.1.1 *Note that homogenous elastic energy densities of the type*

$$W^{\text{el}}(c, e, z) = \frac{1}{2} g(z)\mathbf{C}(e - e^\star(c)) : (e - e^\star(c))$$

with linear eigenstrain $e^\star$ (see (3.6)) and with positive definite and symmetric stiffness tensor $\mathbf{C}$ are covered within the growth assumptions (7.4).

The mobility tensor is assumed to be a scalar non-negative function $M \in \mathcal{C}([0,1]; \mathbb{R}_+)$ which depends on the damage variable and satisfies the degeneracy condition

$$M(z) = 0 \text{ if and only if } z = 0. \tag{7.5}$$

In the next section, we provide a weak formulation of the PDE system introduced in Definition 3.4.3.

7.2 Weak formulations and existence results

As in the previous chapters, the weak formulation will be based on an energetic approach and uses the associated free energy $\mathcal{E}$. Let $G \subseteq \overline{\Omega}$ be a relatively open subset. Then, the free energy contained in the subset G is given by the integral

$$\mathcal{E}_G(c, e, z) := \int_G \left(\frac{1}{p}|\nabla z|^p + \frac{1}{2}|\nabla c|^2 + W^{\mathrm{ch}}(c) + W^{\mathrm{el}}(c, e, z) + f(z) + I_{[0,\infty)}(z)\right) \mathrm{d}x$$

for $c \in H^1(G)$, $e \in L^2(G; \mathbb{R}^{n \times n}_{\mathrm{sym}})$ and $z \in W^{1,p}(G)$. We will omit the subscript G in $\mathcal{E}_G$ and simply write $\mathcal{E}$.

In the following, a weak formulation of the system in Definition 3.4.3 combining the ideas in Section 4.2 and Section 6.2 is given.

Definition 7.2.1 (Weak solution for the system (3.29)-(3.30))
A quadruple (c, u, z, μ) is called a weak solution of the system given in Definition 3.4.3 with the initial-boundary data (c^0, z^0, b) if

(i) *The functions are in the following spaces:*

$$c \in L^\infty(0, T; H^1(\Omega)) \cap H^1(0, T; (H^1(\Omega))^*), \qquad u \in L^2_t H^1_{x,\mathrm{loc}}(F; \mathbb{R}^n),$$
$$z \in L^\infty(0, T; W^{1,p}(\Omega)) \cap SBV^2(0, T; L^2(\Omega)), \qquad \mu \in L^2_t H^1_{x,\mathrm{loc}}(F)$$

with $e := \epsilon(u) \in L^2(F; \mathbb{R}^{n \times n}_{\mathrm{sym}})$ where $F := \mathfrak{A}_{\Gamma_{\mathrm{D}}}(\{z^- > 0\}) \subseteq \overline{\Omega_T}$ is a shrinking set. (Note that z^- denotes the limit from the left side w.r.t. the time variable of the BV function z; see Section 2.2.)

(ii) *Quasi-static mechanical equilibrium:*

$$0 = \int_{F(t)} \partial_e W^{\mathrm{el}}(c(t), e(t), z(t)) : \epsilon(\zeta) \, \mathrm{d}x \tag{7.6}$$

for a.e. $t \in (0, T)$ and for all $\zeta \in H^1_D(\Omega; \mathbb{R}^n)$. Furthermore, $u = b$ on $(\Gamma_{\mathrm{D}})_T \cap F$.

(iii) *Diffusion:*

$$\int_{\Omega_T} \partial_t \zeta(c - c^0) \, \mathrm{d}x \, \mathrm{d}t = \int_F M(z) \nabla \mu \cdot \nabla \zeta \, \mathrm{d}x \, \mathrm{d}t \tag{7.7}$$

for all $\zeta \in L^2(0, T; H^1(\Omega)) \cap H^1(0, T; L^2(\Omega))$ with $\zeta(T) = 0$ and

$$\int_F \mu\zeta \, dx = \int_F \left(\nabla c \cdot \nabla\zeta + \partial_c W^{\mathrm{ch}}(c)\zeta + \partial_c W^{\mathrm{el}}(c, e, z)\zeta\right) dx \tag{7.8}$$

for all $\zeta \in L^2(0, T; H^1(\Omega))$ with $\mathrm{supp}(\zeta) \subseteq F$.

(iv) Damage variational inequality:

$$0 \leq \int_{F(t)} \left(|\nabla z(t)|^{p-2}\nabla z(t) \cdot \nabla\zeta + \left(\partial_z W^{\mathrm{el}}(c(t), e(t), z(t)) + f'(z(t)) + \partial_t^a z(t))\right)\right)\zeta \, dx \tag{7.9}$$

$0 \leq z(t)$ *in* Ω,

$0 \geq \partial_t z(t)$ *a.e. in* Ω

for a.e. $t \in (0, T)$ and for all $\zeta \in W_-^{1,p}(\Omega)$. The initial value is given by $z^+(0) = z^0$.

(v) Damage jump condition:

$$z^+(t) = z^-(t)\mathbb{1}_{F(t)} \text{ in } \overline{\Omega} \tag{7.10}$$

for all $t \in [0, T]$.

(vi) Weak energy inequality:

$$\mathcal{E}(c(t), e(t), z(t)) + \int_0^t \int_{F(s)} \left(|\partial_t^a z|^2 + \mathbb{M}(z)|\nabla\mu|^2\right) dx \, ds + \sum_{s \in J_z \cap (0,t]} \mathfrak{J}_s$$

$$\leq \mathfrak{e}_0^+ + \int_0^t \int_{F(s)} \partial_e W^{\mathrm{el}}(c, e, z) : \epsilon(\partial_t b) \, dx \, ds \tag{7.11}$$

for a.e. $t \in (0, T)$, where the jump part $\mathfrak{J}_s$ satisfies $0 \leq \mathfrak{J}_s$ and is given by

$$\mathfrak{J}_s := \lim_{\tau \to s^-} \operatorname*{ess\,inf}_{\vartheta \in (\tau, s)} \mathcal{E}(c(\vartheta), e(\vartheta), z(\vartheta)) - \mathfrak{e}_s^+ \tag{7.12}$$

and the values $\mathfrak{e}_s^+ \geq 0$ satisfy the upper energy estimate

$$0 \leq \mathfrak{e}_s^+ \leq \mathcal{E}(c(s), \epsilon(b(s) + \zeta), z^+(s)) \tag{7.13}$$

for all $\zeta \in H^1_{\Gamma_D \cap F(s)}(F(s); \mathbb{R}^n)$.

Remark 7.2.2 *Under additional regularity assumptions, a weak solution reduces to the pointwise classical notion given in Definition 3.4.3 (cf. Theorem 6.2.3).*

One aim of this chapter is to prove maximal local-in-time existence of weak solutions according to Definition 7.2.1. In addition, following the approach in Chapter 6, existence of global solutions can be shown in an approximate sense. To be more precise, we use the notation

$$F \approx_\eta \mathfrak{A}_{\Gamma_\mathrm{D}}(\{z^- > 0\})$$

for a measurable set $F \subseteq \overline{\Omega_T}$, a function $z \in SBV^2(0, T; L^2(\Omega))$ and a constant $\eta > 0$ if the conditions

$$F(t) \supseteq \mathfrak{A}_{\Gamma_\mathrm{D}}(\{z^-(t) > 0\}) \text{ for all } t \in [0, T],$$

$$F(t) = \mathfrak{A}_{\Gamma_\mathrm{D}}(\{z^-(t) > 0\}) \text{ for all } t \in [0, T] \setminus \bigcup_{t \in C_{z^\star}} [t, t + \eta),$$

$$\mathcal{L}^n\big(F(t) \setminus \mathfrak{A}_{\Gamma_\mathrm{D}}(\{z^-(t) > 0\})\big) < \eta \text{ for all } t \in \bigcup_{t \in C_{z^\star}} [t, t + \eta)$$

are satisfied. Here, $C_{z^\star}$ denotes the set of cluster points from the right of the jump set $J_{z^\star}$ of the function $z^\star \in SBV^2(0, T; L^2(\Omega))$ given by $z^\star(t) := z(t) \mathbb{1}_{\mathfrak{A}_{\Gamma_\mathrm{D}}(\{z^-(t) > 0\})}$ ($\mathbb{1}_A : X \to \{0, 1\}$ is the characteristic function of a set $A \subseteq X$). Roughly speaking, $z^\star$ is the restricted damage profile of z which takes *all* material exclusions into account.

Definition 7.2.3 (Approximate weak solution for the system (3.29)-(3.30))
A tuple (c, e, u, z, μ) and a shrinking set $F \subseteq \overline{\Omega_T}$ is called an approximate weak solution with fineness $\eta > 0$ of the system given in Definition 3.4.3 with the initial-boundary data (c^0, z^0, b) if

$$c \in L^\infty(0, T; H^1(\Omega)) \cap H^1(0, T; (H^1(\Omega))^*), \qquad u \in L^2_t H^1_{x,\mathrm{loc}}(\mathfrak{A}_{\Gamma_\mathrm{D}}(F); \mathbb{R}^n),$$

$$z \in L^\infty(0, T; W^{1,p}(\Omega)) \cap SBV^2(0, T; L^2(\Omega)), \quad \mu \in L^2_t H^1_{x,\mathrm{loc}}(F),$$

$$e \in L^2(F; \mathbb{R}^{n \times n}_{\mathrm{sym}})$$

with $e = \epsilon(u)$ in $\mathfrak{A}_{\Gamma_\mathrm{D}}(F)$, $F \approx_\eta \mathfrak{A}_{\Gamma_\mathrm{D}}(\{z^- > 0\})$ and properties (ii)-(vi) of Definition 7.2.1 are satisfied.

Theorem 7.2.4 (Global-in-time approximate weak solutions)
Let the assumptions in Section 7.1 be satisfied. Let $b \in W^{1,1}(0, T; W^{1,\infty}(\Omega; \mathbb{R}^n))$, $c^0 \in H^1(\Omega)$ and $z^0 \in W^{1,p}(\Omega)$ with $0 \le z^0 \le 1$ in Ω and $\{z^0 > 0\}$ admissible with respect to Γ_D be initial-boundary data. Furthermore, let $\eta > 0$. Then there exists an approximate weak solution (c, e, u, z, μ) with fineness $\eta > 0$ according to Definition 7.2.3.

Theorem 7.2.5 (Maximal local-in-time existence of weak solutions)
Let the assumptions in Section 7.1 be satisfied. Let $b \in W^{1,1}(0, T; W^{1,\infty}(\Omega; \mathbb{R}^n))$, $c^0 \in H^1(\Omega)$ and $z^0 \in W^{1,p}(\Omega)$ with $0 < \kappa \le z^0 \le 1$ in Ω be initial-boundary data. Then there exist a maximal value $\widehat{T} > 0$ with $\widehat{T} \le T$ and functions c, u, z, μ defined on the time interval $[0, \widehat{T}]$ such that (c, u, z, μ) is a weak solution according to Definition 7.2.1. Therefore, if $\widehat{T} < T$, (c, u, z, μ) cannot be extended to a weak solution on $[0, \widehat{T} + \varepsilon]$ for an $\varepsilon > 0$.

The remaining part of this chapter is devoted to the proof of the local and global existence result.

7.3 Proofs of the existence theorems

7.3.1 Γ-limit of the regularized energy

For each $\delta > 0$, we define the regularized energies $\mathcal{E}_\delta$ and $\mathcal{F}_\delta$ as

$$\mathcal{E}_\delta(c, e, z) := \int_\Omega \left(\frac{1}{p}|\nabla z|^p + \frac{1}{2}|\nabla c|^2 + W^{\mathrm{ch}}(c) + W_\delta^{\mathrm{el}}(c, e, z) + f(z) + I_{[0,\infty)}(z) \right) \mathrm{d}x,$$

$$\mathcal{F}_\delta(c, e, z) := \int_\Omega \left(W_\delta^{\mathrm{el}}(c, e, z) + I_{[0,\infty)}(z) \right) \mathrm{d}x$$

for functions $c \in H^1(\Omega)$, $e \in L^2(\Omega; \mathbb{R}^{n \times n}_{\mathrm{sym}})$ and $z \in W^{1,p}(\Omega)$. The regularized elastic energy density and mobility are given by

$$W_\delta^{\mathrm{el}}(c, e, z) := (g(z) + \delta)\varphi(c, e),$$
$$\mathbb{M}^\delta(z) := \mathbb{M}(z) + \delta.$$

As in the previous chapter, we choose a sequence $\delta_k \to 0^+$ as $k \to \infty$ for the limit passage and omit the subscript k.

We will employ the Γ-limit of the reduced energy functional of $\mathcal{E}_\delta$ in order to gain a suitable energy estimate in the limit $\delta \to 0^+$. To this end, define the reduced energy functionals $\mathfrak{E}_\delta$ and $\mathfrak{F}_\delta$ by

$$\mathfrak{E}_\delta(c, \xi, z) := \min_{\zeta \in H^1_{\Gamma_{\mathrm{D}}}(\Omega; \mathbb{R}^n)} \mathcal{E}_\delta(c, \epsilon(\xi + \zeta), z),$$

$$\mathfrak{F}_\delta(c, \xi, z) := \min_{\zeta \in H^1_{\Gamma_{\mathrm{D}}}(\Omega; \mathbb{R}^n)} \mathcal{F}_\delta(c, \epsilon(\xi + \zeta), z).$$

The Γ-limits of $\mathfrak{E}_\delta$ and $\mathfrak{F}_\delta$ as $\delta \to 0^+$ exist in the topological space $H^1_{\mathrm{w}}(\Omega) \times W^{1,\infty}(\Omega; \mathbb{R}^n) \times W^{1,p}_{\mathrm{w}}(\Omega)$ and are denoted by $\mathfrak{E}$ and $\mathfrak{F}$, respectively. The limit functional $\mathfrak{F}$ is needed as an auxiliary construction in the following because it already captures the essential properties of $\mathfrak{E}$. In the next section, we are going to prove some properties of the Γ-limit $\mathfrak{E}$ which are used in the global-in-time existence proof.

The proof of global-in-time existence of approximate weak solutions requires a concatenation property (see Lemma 7.3.9) which is, in turn, based on some deeper insights into the Γ-limit $\mathfrak{E}$. To this end, it is necessary to have more information about the recovery sequences for $\mathfrak{F}_\delta \xrightarrow{\Gamma} \mathfrak{F}$.

To proceed, we will introduce the following substitution method. Assume that $u \in H^1(\Omega; \mathbb{R}^n)$ minimizes $\mathcal{F}_\delta(c, \epsilon(\cdot), z)$ with Dirichlet data ξ on Γ_{D}. Then by expressing the elastic energy density W^{el} in terms of its derivative $\partial_e W^{\mathrm{el}}$, i.e.,

$$W^{\mathrm{el}} = \frac{1}{2}\partial_e W^{\mathrm{el}} : e + \frac{1}{2}z\varphi^2(c) : e + z\varphi^3(c),$$

and by testing the Euler-Lagrange equation for u with $\zeta = u - \tilde{u}$ for a function $\tilde{u} \in H^1(\Omega; \mathbb{R}^n)$ with $\tilde{u} = \xi$ on Γ_{D}, the elastic energy term in $\mathcal{F}_\delta$ can be rewritten as

$$\int_\Omega W_\delta^{\mathrm{el}}(c, \epsilon(u), z) \, \mathrm{d}x = \int_\Omega (g(z) + \delta) \left(\varphi^1 \epsilon(u) : \epsilon(\tilde{u}) + \frac{1}{2}\varphi^2(c) : (\epsilon(u) + \epsilon(\tilde{u})) + \varphi^3(c) \right) \mathrm{d}x.$$
$$(7.14)$$

For convenience, in the following proof, the density $\widetilde{W}_\delta^{\mathrm{el}}$ is defined as

$$\widetilde{W}_\delta^{\mathrm{el}}(c, e, e_1, z) := (g(z) + \delta)\left(\varphi^1 e : e_1 + \frac{1}{2}\varphi^2(c) : (e + e_1) + \varphi^3(c)\right).$$

Lemma 7.3.1 *For every $c \in H^1(\Omega)$, $\xi \in W^{1,\infty}(\Omega)$ and $z \in W_+^{1,p}(\Omega)$ there exists a sequence $\lambda_\delta \to 0^+$ such that $(c, \xi, (z-\lambda_\delta)^+) \to (c, \xi, z)$ is a recovery sequence for $\mathfrak{F}_\delta \xrightarrow{\Gamma} \mathfrak{F}$.*

Proof. We follow the idea of the proof in Lemma 6.3.3. But here we have the additional concentration variable c which complicates the calculation. Let $(c_\delta, \xi_\delta, z_\delta) \to (c, \xi, z)$ be a recovery sequence such that $(z - \lambda_\delta)^+ \leq z_\delta$ for some sequence $\lambda_\delta \to 0^+$. Consider

$$\mathfrak{F}_\delta(c, \xi, (z - \lambda_\delta)^+) - \mathfrak{F}_\delta(c_\delta, \xi_\delta, z_\delta)$$
$$= \underbrace{\mathfrak{F}_\delta(c, \xi, (z - \lambda_\delta)^+) - \mathfrak{F}_\delta(c, \xi, z_\delta)}_{A_\delta} + \underbrace{\mathfrak{F}_\delta(c, \xi, z_\delta) - \mathfrak{F}_\delta(c_\delta, \xi_\delta, z_\delta)}_{B_\delta}.$$

Since $A_\delta \leq 0$, we focus on the second term of the right hand side. Let $u_\delta, v_\delta \in H^1_{\Gamma_D}(\Omega; \mathbb{R}^n)$ be given by

$$u_\delta = \underset{\zeta \in H^1_{\Gamma_D}(\Omega;\mathbb{R}^n)}{\arg\min} \mathfrak{F}_\delta(c, \epsilon(\xi + \zeta), z_\delta), \quad v_\delta = \underset{\zeta \in H^1_{\Gamma_D}(\Omega;\mathbb{R}^n)}{\arg\min} \mathfrak{F}_\delta(c_\delta, \epsilon(\xi_\delta + \zeta), z_\delta).$$

Using (7.14) for $(c, \xi + u_\delta, z_\delta)$ with test function $\tilde{u} = v_\delta$ and (7.14) for $(c_\delta, \xi_\delta + v_\delta, z_\delta)$ with test function $\tilde{u} = u_\delta$, we obtain a calculation as follows:

$$B_\delta = \mathfrak{F}_\delta(c, \epsilon(\xi + u_\delta), z_\delta) - \mathfrak{F}_\delta(c_\delta, \epsilon(\xi_\delta + v_\delta), z_\delta,)$$
$$= \int_\Omega \left(\widetilde{W}^{\mathrm{el}}(c, \epsilon(\xi + u_\delta), \epsilon(\xi + v_\delta), z_\delta + \delta) - \widetilde{W}^{\mathrm{el}}(c_\delta, \epsilon(\xi_\delta + v_\delta), \epsilon(\xi_\delta + u_\delta), z_\delta + \delta)\right) dx$$
$$= \int_\Omega (g(z_\delta) + \delta)\left(\varphi^1 \epsilon(\xi + u_\delta) : \epsilon(\xi + v_\delta) - \varphi^1 \epsilon(\xi_\delta + u_\delta) : \epsilon(\xi_\delta + v_\delta)\right.$$
$$\left. + \frac{1}{2}\varphi^2(c) : \epsilon(2\xi + u_\delta + v_\delta) - \frac{1}{2}\varphi^2(c_\delta) : \epsilon(2\xi_\delta + v_\delta + u_\delta) + \varphi^3(c) - \varphi^3(c_\delta)\right) dx$$
$$= \int_\Omega (g(z_\delta) + \delta)\left(\varphi^1 \epsilon(\xi) : \epsilon(\xi) - \varphi^1 \epsilon(\xi_\delta) : \epsilon(\xi_\delta) + \varphi^1 \epsilon(u_\delta + v_\delta) : \epsilon(\xi - \xi_\delta)\right.$$
$$\left. + \varphi^2(c) : \epsilon(\xi - \xi_\delta) + \frac{1}{2}(\varphi^2(c) - \varphi^2(c_\delta)) : \epsilon(2\xi_\delta + u_\delta + v_\delta) + \varphi^3(c) - \varphi^3(c_\delta)\right) dx$$
$$\leq \int_\Omega (g(z_\delta) + \delta)\left(\varphi^1 \epsilon(\xi) : \epsilon(\xi) - \varphi^1 \epsilon(\xi_\delta) : \epsilon(\xi_\delta) + \varphi^2(c) : \epsilon(\xi - \xi_\delta) + \varphi^3(c) - \varphi^3(c_\delta)\right) dx$$
$$+ \|(g(z_\delta) + \delta)\varphi^1 \epsilon(u_\delta + v_\delta)\|_{L^2(\Omega)} \|\epsilon(\xi - \xi_\delta)\|_{L^2(\Omega)}$$
$$+ \frac{1}{2}\|\varphi^2(c) - \varphi^2(c_\delta)\|_{L^2(\Omega)}\left(\|(g(z_\delta) + \delta)\epsilon(\xi_\delta + u_\delta)\|_{L^2(\Omega)} + \|(g(z_\delta) + \delta)\epsilon(\xi_\delta + v_\delta)\|_{L^2(\Omega)}\right).$$

Using the convergence properties $c_\delta \rightharpoonup c$ in $H^1(\Omega)$, $\xi_\delta \to \xi$ in $W^{1,\infty}(\Omega)$, $z_\delta \rightharpoonup z$ in $W^{1,p}(\Omega)$ and the boundedness of $\mathfrak{F}_\delta(c, \epsilon(\xi + u_\delta), z_\delta)$ and $\mathfrak{F}_\delta(c_\delta, \epsilon(\xi_\delta + v_\delta), z_\delta)$ with respect to δ, we conclude $\limsup_{\delta \to 0^+} B_\delta \leq 0$. The claim follows as in Lemma 6.3.3. $\square$

Remark 7.3.2 *The knowledge of the recovery sequences for $\mathfrak{F}_\delta \xrightarrow{\Gamma} \mathfrak{F}$ gives also more information about $\mathfrak{E}$. In particular, we obtain an analogous result for $\mathfrak{E}$ as in Lemma 7.3.1 and, moreover, the following properties (cf. Corollary 6.3.4 and Lemma 6.3.6]):*

- $\mathfrak{E}(c, \xi, \mathbb{1}_F z) \leq \mathfrak{E}(c, \xi, z)$ $\forall c \in H^1(\Omega),\ \forall \xi \in W^{1,\infty}(\Omega; \mathbb{R}^n),\ \forall z \in W^{1,p}(\Omega)$ $\forall F \subseteq \Omega$ *open with* $\mathbb{1}_F z \in W^{1,p}(\Omega)$,

- $\mathfrak{E}(c, \xi, z) \leq \mathcal{E}(c, \epsilon(u), z)$ $\forall c \in H^1(\Omega),\ \forall \xi \in W^{1,\infty}(\Omega; \mathbb{R}^n),$ $\forall z \in W^{1,p}(\Omega)$ *with* $0 \leq z \leq 1,$ $\forall u \in H^1_{\mathrm{loc}}(\{z > 0\}; \mathbb{R}^n)$ *with* $u = \xi$ *on* $D \cap \{z > 0\}$.

7.3.2   Degenerate limit of the regularized system

In this section, we will review the corresponding incomplete damage model coupled to an elastic Cahn-Hilliard system and then perform a degenerate limit procedure.

A modification of the proof of Theorem 4.2.7 yields the following existence result for system (3.27)-(3.28) with damage dependent mobility.

Theorem 7.3.3 (δ-regularized coupled PDE system) *Let $\delta > 0$. For given initial-boundary data $c^0_\delta \in H^1(\Omega)$, $z^0_\delta \in W^{1,p}(\Omega)$ and $b_\delta \in W^{1,1}(0, T; W^{1,\infty}(\Omega; \mathbb{R}^n))$ there exists a quadruple $(c_\delta, u_\delta, z_\delta, \mu_\delta)$ such that*

(i) *The functions are in the following spaces:*

$$c_\delta \in L^\infty(0, T; H^1(\Omega)) \cap H^1(0, T; (H^1(\Omega))^*), \quad u_\delta \in L^\infty(0, T; H^1(\Omega; \mathbb{R}^n)),$$
$$z_\delta \in L^\infty(0, T; W^{1,p}(\Omega)) \cap H^1(0, T; L^2(\Omega)), \quad \mu_\delta \in L^2(0, T; H^1(\Omega)).$$

(ii) *Quasi-static mechanical equilibrium:*

$$\int_\Omega \partial_e W^{\mathrm{el}}_\delta(c_\delta(t), \epsilon(u_\delta(t)), z_\delta(t)) : \epsilon(\zeta)\, \mathrm{d}x = 0 \tag{7.15}$$

for a.e. $t \in (0, T)$ and for all $\zeta \in H^1_{\Gamma_D}(\Omega; \mathbb{R}^n)$. Furthermore, $u_\delta = b_\delta$ on the boundary $(\Gamma_D)_T$.

(iii) *Diffusion:*

$$\int_{\Omega_T} (c_\delta - c^0_\delta) \partial_t \zeta\, \mathrm{d}x\, \mathrm{d}t = \int_{\Omega_T} \mathbb{M}^\delta(z_\delta) \nabla \mu_\delta \cdot \nabla \zeta\, \mathrm{d}x\, \mathrm{d}t \tag{7.16}$$

for all $\zeta \in L^2(0, T; H^1(\Omega))$ with $\partial_t \zeta \in L^2(\Omega_T)$ and $\zeta(T) = 0$ and

$$\int_\Omega \mu_\delta(t)\zeta\, \mathrm{d}x = \int_\Omega \left(\nabla c_\delta(t) \cdot \nabla \zeta + \partial_c W^{\mathrm{ch}}(c_\delta(t))\zeta + \partial_c W^{\mathrm{el}}_\delta(c_\delta(t), \epsilon(u_\delta(t)), z_\delta(t))\zeta \right) \mathrm{d}x \tag{7.17}$$

for a.e. $t \in (0, T)$ and for all $\zeta \in H^1(\Omega)$.

(iv) Damage variational inequality:

$$0 \leq \int_\Omega |\nabla z_\delta(t)|^{p-2} \nabla z_\delta(t) \cdot \nabla \zeta \, dx$$
$$+ \int_\Omega \left(\partial_z W_\delta^{\mathrm{el}}(c_\delta(t), \epsilon(u_\delta(t)), z_\delta(t)) + f'(z_\delta(t)) + \partial_t z_\delta(t) + r_\delta(t) \right) \zeta \Big) dx \quad (7.18)$$
$$0 \leq z_\delta(t) \text{ in } \Omega,$$
$$0 \geq \partial_t z_\delta(t) \text{ a.e. in } \Omega$$

for a.e. $t \in (0,T)$ and for all $\zeta \in W^{1,p}(\Omega)$ with $\zeta \leq 0$ where $r_\delta \in L^1(\Omega_T)$ satisfies

$$r_\delta = -\chi_\delta \left[\partial_z W_\delta^{\mathrm{el}}(c_\delta, \epsilon(u_\delta), z_\delta) + f'(z) \right]^+ \quad (7.19)$$

with $\chi_\delta \in L^\infty(\Omega)$ fulfilling $\chi_\delta = 0$ on $\{z_\delta > 0\}$ and $0 \leq \chi_\delta \leq 1$ on $\{z_\delta = 0\}$. The initial value is given by $z_\delta(0) = z_\delta^0$.

(v) Energy inequality:

$$\mathcal{E}_\delta(c_\delta(t), \epsilon(u_\delta(t)), z_\delta(t)) + \int_{\Omega_t} \left(|\partial_t z_\delta|^2 + \mathbb{M}^\delta(z_\delta)|\nabla \mu_\delta|^2 \right) dx \, ds$$
$$\leq \mathcal{E}_\delta(c_\delta^0, \epsilon(u_\delta^0), z_\delta^0) + \int_{\Omega_t} \partial_e W^{\mathrm{el}}(c_\delta, \epsilon(u_\delta), z_\delta) : \epsilon(\partial_t b_\delta) \, dx \, ds \quad (7.20)$$

holds for a.e. $t \in (0,T)$ where u_δ^0 minimizes $\mathcal{E}_\delta(c^0, \epsilon(\cdot), z_\delta^0)$ in $H^1(\Omega; \mathbb{R}^n)$ with Dirichlet data $b_\delta^0 := b_\delta(0)$ on Γ_D.

Proof. The existence theorem presented in Theorem 4.2.7 can be adapted to our situation by considering the viscous semi-implicit time-discretized system (in a classical notation; we omit the δ-dependence in the notation for the discrete solution at the moment)

$$0 = \mathrm{div}\left(\partial_e W_\delta^{\mathrm{el}}(c^m, \epsilon(u^m), z^m) \right) + \lambda \, \mathrm{div}(|u^m|^2 u^m),$$
$$\frac{c^m - c^{m-1}}{\tau} = \mathrm{div}(\mathbb{M}^\delta(z^{m-1})\nabla \mu^m),$$
$$\mu^m = -\Delta c^m + \partial_c W^{\mathrm{ch}}(c^m) + \partial_c W_\delta^{\mathrm{el}}(c^m, \epsilon(u^m), z^m) + \lambda \frac{c^m - c^{m-1}}{\tau},$$
$$\frac{z^m - z^{m-1}}{\tau} + \zeta + \varrho = \mathrm{div}(|\nabla z^m|^{p-2}\nabla z^m) + \partial_z W_\delta^{\mathrm{el}}(c^m, \epsilon(u^m), z^m) + f'(z^m),$$

with the subgradients $\zeta \in \partial I_{[0,\infty)}(z^m)$, $\varrho \in \partial I_{(-\infty,0]}((z^m - z^{m-1})/\tau)$ and the discretization fineness $\tau = T/M$ for $M \in \mathbb{N}$. The discrete equations can be obtained recursively starting from (c^0, u^0, z^0) with $u^0 := \arg\min_{u \in H^1(\Omega;\mathbb{R}^n),\, u|_{\Gamma_D} = b^0|_{\Gamma_D}} \mathcal{E}_\delta(c^0, u, z^0)$ by considering the Euler-Lagrange equations of the functional

$$\mathbb{E}^m(c, u, z) := \mathcal{E}_\delta(c, u, z) + \int_\Omega \frac{\lambda}{4} |\nabla u|^4 \, dx$$
$$+ \frac{\tau}{2} \left(\left\| \frac{z - z^{m-1}}{\tau} \right\|_{L^2(\Omega)}^2 + \left\| \frac{c - c^{m-1}}{\tau} \right\|_{X(z^{m-1})}^2 + \lambda \left\| \frac{c - c^{m-1}}{\tau} \right\|_{L^2(\Omega)}^2 \right)$$

defined on the subspace of $H^1(\Omega) \times W^{1,4}(\Omega; \mathbb{R}^n) \times W^{1,p}(\Omega)$ given by the conditions $u|_{\Gamma_D} = b(m\tau)|_{\Gamma_D}$, $\int_\Omega (c - c^0) \, dx = 0$ and $0 \leq z \leq z^{m-1}$ a.e. in Ω. The scalar product $\langle \cdot, \cdot \rangle_{X(z^{m-1})}$ is given by

$$\langle u, v \rangle_{X(z^{m-1})} := \left\langle \mathbb{M}^\delta(z^{m-1}) \nabla A^{-1} u, \nabla A^{-1} v \right\rangle_{L^2(\Omega)}$$

with the operator $A : V_0 \to \widetilde{V}_0$, $Au := \left\langle \mathbb{M}^\delta(z^{m-1}) \nabla u, \nabla \cdot \right\rangle_{L^2(\Omega)}$ and the spaces V_0 and $\widetilde{V}_0$ given in (4.23).

After passing the discretization fineness τ to 0 and, then, passing $\lambda \to 0^+$, we obtain the equations and inequalities (7.15)-(7.20). Note that the mobility $\mathbb{M}^\delta(z_M)$ is uniformly bounded from below w.r.t. M by a positive constant and converges uniformly to $\mathbb{M}(z_\delta)$. $\square$

Let $(c_\delta^0, b_\delta^0, z_\delta^0) \to (c^0, b^0, z^0)$ as $\delta \to 0^+$ be a recovery sequence for $\mathfrak{E}_\delta \xrightarrow{\Gamma} \mathfrak{E}$. In particular, $c_\delta^0 \rightharpoonup c^0$ in $H^1(\Omega)$, $b_\delta^0 \to b^0$ in $W^{1,\infty}(\Omega; \mathbb{R}^n)$ and $z_\delta^0 \rightharpoonup z^0$ in $W^{1,p}(\Omega)$. Furthermore, we set $b_\delta := b - b^0 + b_\delta^0$. For each $\delta > 0$, we obtain a weak solution $(c_\delta, u_\delta, z_\delta, \mu_\delta)$ for $(c_\delta^0, z_\delta^0, b_\delta)$ according to Theorem 7.3.3.

Applying Gronwall's lemma to the energy estimate (7.20) and following the argumentation in Lemma 6.3.10 for the variables $\widehat{e}_\delta$ and z_δ, we gain the following a-priori estimates:

- $\sup_{t \in [0,T]} \|c_\delta(t)\|_{H^1(\Omega)} \leq C$,
- $\|\widehat{e}_\delta\|_{L^2(\Omega_T; \mathbb{R}^{n \times n})} \leq C$
  with $\widehat{e}_\delta := e_\delta \mathbb{1}_{\{z_\delta > 0\}}$,
- $\sup_{t \in [0,T]} \|z_\delta(t)\|_{W^{1,p}(\Omega)} \leq C$,
- $\|\partial_t z_\delta\|_{L^2(\Omega_T)} \leq C$,

- $\|W_\delta^{\text{el}}(c_\delta, e_\delta, z_\delta)\|_{L^\infty(0,T;L^1(\Omega))} \leq C$,
- $\|\mathbb{M}^\delta(z_\delta)^{1/2} \nabla \mu_\delta\|_{L^2(\Omega_T; \mathbb{R}^n)} \leq C$,
- $\|\partial_t c_\delta\|_{L^2(0,T;(H^1(\Omega))^*)} \leq \|\mathbb{M}^\delta(z_\delta) \nabla \mu_\delta\|_{L^2(\Omega_T; \mathbb{R}^n)}$
  $\leq C$.

These estimates, an Aubin-Lions type compactness theorem (see Theorem 2.3.9 (i)), the variational inequality (7.18) and an approximation argument (see Lemma 2.3.18) yield the following convergence properties (cf. Lemma 6.3.11 for details):

Lemma 7.3.4 *There exists functions*

(i) $c \in L^\infty(0,T;H^1(\Omega))$
     $\cap H^1(0,T;(H^1(\Omega))^*)$,

(ii) $\widehat{e} \in L^2(\Omega_T; \mathbb{R}^{n \times n})$,

(iii) $z \in L^\infty(0,T;W^{1,p}(\Omega))$
     $\cap H^1(0,T;L^2(\Omega))$,
     z is monotonically decreasing
     with respect to t, i.e., $\partial_t z \leq 0$

and a subsequence (we omit the index) such that for $\delta \to 0^+$

(a) $c_\delta \rightharpoonup c$ in $H^1(0,T;(H^1(\Omega))^*)$,
$c_\delta \to c$ in $L^r(\Omega_T)$ for all $1 \le r < 2^\star$,
$c_\delta(t) \rightharpoonup c(t)$ in $H^1(\Omega)$ for all t,
$c_\delta \to c$ a.e. in Ω_T,

(b) $z_\delta \rightharpoonup z$ in $H^1(0,T;L^2(\Omega))$,
$z_\delta \to z$ in $L^p(0,T;W^{1,p}(\Omega))$,
$z_\delta(t) \rightharpoonup z(t)$ in $W^{1,p}(\Omega)$ for all t,
$z_\delta \to z$ in $\overline{\Omega_T}$,

(c) $b_\delta \to b$ in $W^{1,1}(0,T;W^{1,\infty}(\Omega;\mathbb{R}^n))$,

(d) $\widehat{e}_\delta \rightharpoonup \widehat{e}$ in $L^2(\Omega_T;\mathbb{R}^{n\times n})$,
$\partial_e W_\delta^{\mathrm{el}}(c_\delta,e_\delta,z_\delta) \rightharpoonup \partial_e W^{\mathrm{el}}(c,\widehat{e},z)$
in $L^2(\{z>0\};\mathbb{R}^{n\times n})$,
$\partial_e W_\delta^{\mathrm{el}}(c_\delta,e_\delta,z_\delta) \to 0$
in $L^2(\{z=0\};\mathbb{R}^{n\times n})$,
$\partial_c W_\delta^{\mathrm{el}}(c_\delta,e_\delta,z_\delta) \rightharpoonup \partial_c W^{\mathrm{el}}(c,\widehat{e},z)$
in $L^2(\{z>0\};\mathbb{R}^{n\times n})$,
$\partial_c W_\delta^{\mathrm{el}}(c_\delta,e_\delta,z_\delta) \to 0$
in $L^2(\{z=0\};\mathbb{R}^{n\times n})$.

Lemma 7.3.5 (A-priori estimates for μ_δ)

(i) *Interior estimate.* For every $t \in [0,T]$ and for every open cube $Q \subset\subset \{z(t) > 0\} \cap \Omega$, there exists a $C > 0$ such that for all sufficiently small $\delta > 0$ (abbr. $0 < \delta \ll 1$)

$$\|\mu_\delta\|_{L^2(0,t;H^1(Q))} \le C. \tag{7.21}$$

(ii) *Estimate at the boundary.* For every $t \in [0,T]$ and every $x_0 \in \{z(t) > 0\} \cap \partial\Omega$, there exist a neighborhood U of x_0 and a $C > 0$ such that for all $0 < \delta \ll 1$

$$\|\mu_\delta\|_{L^2(0,t;H^1(U\cap\Omega))} \le C. \tag{7.22}$$

Proof.

(i) Let $t \in [0,T]$ and $Q \subset\subset \{z(t) > 0\} \cap \Omega$ be an open cube. We consider the Lipschitz domain $\widetilde{Q} := B_\eta(Q) := \{x \in \mathbb{R}^n \,|\, \mathrm{dist}(x,Q) < \delta\}$, where $\eta > 0$ is chosen so small such that $\widetilde{Q} \subset\subset \{z(t) > 0\} \cap \Omega$. We define the following function

$$\zeta(x) := \begin{cases} \lambda(x) := \mathrm{dist}(x,\partial\widetilde{Q}) & \text{if } x \in \widetilde{Q}, \\ 0 & \text{else.} \end{cases}$$

The function ζ is a Lipschitz function on $\overline{\Omega}$ with Lipschitz constant 1.

- This can be proven elementary. Let $x,y \in \widetilde{Q}$ be arbitrary. Since $\partial\widetilde{Q}$ is closed, there exist $x_0,y_0 \in \partial\widetilde{Q}$ such that $dist(x,\partial\widetilde{Q}) = |x - x_0|$ and $dist(y,\partial\widetilde{Q}) = |y - y_0|$. WLOG $|x - x_0| \ge |y - y_0|$. Consequently, by noticing $|x - x_0| = dist(x,\partial\widetilde{Q}) \le |x - y_0|$,

$$\left| \mathrm{dist}(x,\partial\widetilde{Q}) - \mathrm{dist}(y,\partial\widetilde{Q}) \right| = ||x - x_0| - |y - y_0|| = |x - x_0| - |y_0 - y|$$
$$\le |x - y_0| - |y_0 - y|$$
$$\le |x - y_0 + y_0 - y|$$
$$= |x - y|.$$

By Rademacher's theorem (see, for instance, [Zie89, Theorem 2.2.1]), ζ is in $W^{1,\infty}(\Omega)$. Now, we can test (7.17) with ζ. Then, using the previous a-priori estimates yield boundedness of

$$\int_{\widetilde{Q}} \mu_\delta(x,s)\lambda(x)\,\mathrm{d}x \leq C \qquad (7.23)$$

with respect to a.e. $s \in (0,T)$ and δ.

There exists an $\eta > 0$ such that $z_\delta(s) \geq \eta$ in $\widetilde{Q}$ for all $s \in [0,t]$ and for all $0 < \delta \ll 1$ (see Corollary 6.3.12). Thus, by assumption (7.5), $\mathbb{M}^\delta(z_\delta(s)) \geq \eta' > 0$ holds in $\widetilde{Q}$ for all $s \in [0,t]$ and all $0 < \delta \ll 1$ for a common constant $\eta' > 0$. Consequently, we get by the a-priori estimate for $\mathbb{M}^\delta(z_\delta)^{1/2}\nabla\mu_\delta$

$$\|\nabla\mu_\delta\|_{L^2(\widetilde{Q}\times[0,t])} \leq C \qquad (7.24)$$

for all δ. Applying Theorem 2.3.11 (we plug in $\Omega = \widetilde{Q}$, $r = p = 2$ and $w = \mu_\delta(s)$ for $s \in [0,t]$), integrating from 0 to t and using boundedness properties (7.23) and (7.24), we obtain boundedness of $\|\mu_\delta\lambda\|_{L^2(\widetilde{Q}\times[0,t])}$ and thus boundedness of $\|\mu_\delta\|_{L^2(Q\times[0,t])}$ with respect to $0 < \delta \ll 1$. Together with (7.24), we get the claim (7.21).

(ii) Since $\{z(t) = 0\} \subseteq \overline{\Omega}$ is a closed set, we can find a neighborhood $U \subseteq \mathbb{R}^n\backslash\{z(t) = 0\}$ of x_0. Furthermore, since Ω has a $\mathcal{C}^2$-boundary, there exists a $\mathcal{C}^2$-diffeomorphism $\pi : (-1,1)^n \to U$ with the properties

- $\pi\big((-1,1)^{n-1} \times (-1,0)\big) \subseteq \Omega$,
- $\pi\big((-1,1)^{n-1} \times \{0\}\big) \subseteq \partial\Omega$,
- $\pi\big((-1,1)^{n-1} \times (0,1)\big) \subseteq \mathbb{R}^n \setminus \overline{\Omega}$.

Let $\vartheta : (-1,1)^n \to (-1,1)^n$ denote the reflection $x \mapsto (x_1,\ldots,x_{n-1},-x_n)$ and $\mathcal{T} := \pi \circ \vartheta \circ \pi^{-1}$. Furthermore, let $\widetilde{\mu}_\delta \in L^2(0,t;H^1(U))$ be defined by

$$\widetilde{\mu}_\delta(x,s) := \begin{cases} \mu_\delta(x,s) & \text{if } x \in U \cap \Omega, \\ \mu_\delta(\mathcal{T}(x),s) & \text{if } x \in U \setminus \overline{\Omega}. \end{cases}$$

Let $Q \subset\subset U$ be a non-empty open cube with $x_0 \in Q$. Then, integration by substitution with respect to the transformation $\mathcal{T}$ yields for a.e. $s \in (0,t)$

$$\int_Q \widetilde{\mu}_\delta(x,s)\lambda(x)\,\mathrm{d}x = \int_{Q\cap\Omega} \mu_\delta(x,s)\lambda(x)\,\mathrm{d}x$$
$$+ \int_{\mathcal{T}(Q\backslash\Omega)} \mu_\delta(x,s)\lambda(\mathcal{T}(x))|\det(\nabla\mathcal{T}(x))|\,\mathrm{d}x, \qquad (7.25)$$

where the Lipschitz function $\lambda : \mathbb{R}^N \to \mathbb{R}$ is given by

$$\lambda(x) := \begin{cases} \mathrm{dist}(x,\partial Q) & \text{if } x \in Q, \\ 0 & \text{if } x \in \mathbb{R}^n \setminus Q. \end{cases}$$

We are going to show that both terms on the right hand side of (7.25) are bounded with respect to δ and a.e. $s \in (0, t)$.

- Testing (7.17) with the function $\zeta = \lambda$ yields

$$
\int_{Q \cap \Omega} \mu_\delta(s) \lambda \, dx = \int_{Q \cap \Omega} \left(\nabla c_\delta(s) \cdot \nabla \lambda + \partial_c W^{\text{ch}}(c_\delta(s)) \lambda \right) dx
$$
$$
+ \int_{Q \cap \Omega} \partial_c W_\delta^{\text{el}}(c_\delta(s), \epsilon(u_\delta(s)), z_\delta(s)) \lambda \, dx.
$$

By the already known a-priori estimates, every integral term on the right hand side is bounded w.r.t. δ and a.e. $s \in (0, T)$.

- The function

$$
\zeta(x) := \begin{cases} (\lambda(\mathcal{T}(x))) |\det(\nabla \mathcal{T}(x))| & \text{if } x \in \mathcal{T}(Q \setminus \Omega), \\ 0 & \text{if } x \in \Omega \setminus \mathcal{T}(Q \setminus \Omega) \end{cases}
$$

is a Lipschitz function in Ω because:

  - $\lambda \circ \mathcal{T}$ is a Lipschitz function in $U \cap \Omega$ and $\lambda \circ \mathcal{T} = 0$ in $(U \cap \Omega) \setminus \mathcal{T}(Q \setminus \Omega)$. The first property follows from the Lipschitz continuity of λ and of $\mathcal{T}$ (note that $\mathcal{T}$ is a $\mathcal{C}^2$-diffeomorphism). The latter property can be seen as follows. *Assume the contrary.* Then, we find an $x \in (U \cap \Omega) \setminus \mathcal{T}(Q \setminus \Omega)$ such that $\lambda(\mathcal{T}(x)) > 0$. By the definition of λ, we get $\mathcal{T}(x) \in Q$. Since $x \in \Omega$, it follows $\mathcal{T}(x) \notin \Omega$ by the construction of $\mathcal{T}$. Therefore, $\mathcal{T}(x) \in Q \setminus \Omega$. This gives $x = \mathcal{T}(\mathcal{T}(x)) \in \mathcal{T}(Q \setminus \Omega)$ which is a contradiction.
  - $|\det(\nabla \mathcal{T})|$ is a Lipschitz function in $U \cap \Omega$ ($\mathcal{T}$ is a $\mathcal{C}^2$-diffeomorphism).

Testing (7.17) with ζ yields

$$
\int_{\mathcal{T}(Q \setminus \Omega)} \mu_\delta(s)(\lambda \circ \mathcal{T}) |\det(\nabla \mathcal{T})| \, dx
$$
$$
= \int_{\mathcal{T}(Q \setminus \Omega)} \nabla c_\delta(s) \cdot \nabla \left((\lambda \circ \mathcal{T}) |\det(\nabla \mathcal{T})| \right) dx
$$
$$
+ \int_{\mathcal{T}(Q \setminus \Omega)} \left(\partial_c W^{\text{ch}}(c_\delta(s)) + \partial_c W_\delta^{\text{el}}(c_\delta(s), \epsilon(u_\delta(s)), z_\delta(s)) \right) (\lambda \circ \mathcal{T}) |\det(\nabla \mathcal{T})| \, dx.
$$

By the already known a-priori estimates, every integral term on the right hand side is bounded w.r.t. δ and a.e. $s \in (0, t)$.

For $\nabla \widetilde{\mu}_\delta(s)$, we also get by integration via substitution:

$$
\int_0^t \int_Q |\nabla \widetilde{\mu}_\delta(x, s)|^2 \, dx \, ds
$$
$$
\leq \int_0^t \int_{Q \cap \Omega} |\nabla \mu_\delta(x, s)|^2 \, dx \, ds + \int_0^t \int_{Q \setminus \Omega} |\nabla \mu_\delta(\mathcal{T}(x), s)|^2 |\nabla \mathcal{T}(x)|^2 \, dx \, ds
$$

$$= \int_0^t \int_{Q \cap \Omega} |\nabla \mu_\delta(x, s)|^2 \, dx \, ds$$

$$+ \int_0^t \int_{\mathfrak{I}(Q \setminus \Omega)} |\nabla \mu_\delta(x, s)|^2 |\nabla \mathfrak{I}(\mathfrak{I}(x))|^2| \det(\nabla \mathfrak{I}(x))| \, dx \, ds. \qquad (7.26)$$

Since $Q \cap \Omega \subset\subset \{z(t) > 0\}$ and $\mathfrak{I}(Q \setminus \Omega) \subset\subset \{z(t) > 0\}$, we deduce $z_\delta(s) \geq \eta$ on $Q \cap \Omega$ and on $\mathfrak{I}(Q \setminus \Omega)$ for all $s \in [0, t]$ and for all sufficiently small $0 < \delta$ (see Corollary 6.3.12). Thus, $\nabla \mu_\delta$ is bounded in $L^2((Q \cap \Omega) \times (0, t); \mathbb{R}^n)$ and in $L^2(\mathfrak{I}(Q \setminus \Omega) \times (0, t); \mathbb{R}^n)$ with respect to $0 < \delta \ll 1$ by also using the a-priori estimate for $\mathrm{M}^\delta(z_\delta)^{1/2} \nabla \mu_\delta$, the property $\mathrm{M} \in \mathcal{C}([0, 1]; \mathbb{R}_+)$ and assumption (7.5).

Therefore, the left hand side of (7.26) is also bounded for all $0 < \delta \ll 1$. The Conical Poincaré inequality in Theorem 2.3.11 yields boundedness of $\widetilde{\mu}_\delta \lambda$ in $L^2(Q \times (0, t))$. Finally, we can find a neighborhood $V \subseteq Q$ of x_0 such that $\widetilde{\mu}_\delta$ is bounded in $L^2(0, t; H^1(V))$. $\qquad \square$

Due to the a-priori estimates for $\{u_\delta\}$ and $\{\mu_\delta\}$, the limit functions u and μ can only be expected to be in some space-time local Sobolev space $L_t^2 H_{x,\mathrm{loc}}^1$ (see Subsection 2.4.2). In the sequel, it will be necessary to represent the maximal admissible subset of the not completely damaged area, i.e., $\mathfrak{A}_{\Gamma_\mathrm{D}}(\{z > 0\})$, as a union of Lipschitz domains which capture some parts of the Dirichlet boundary Γ_D. Following the argumentation in Lemma 6.3.13, we define the shrinking set $F := \{z > 0\}$ and obtain the following result.

Lemma 7.3.6 *There exists a function $u \in L_t^2 H_{x,\mathrm{loc}}^1(\mathfrak{A}(F); \mathbb{R}^n)$ such that $\epsilon(u) = \widehat{e}$ a.e. in $\mathfrak{A}_\Gamma(F)$ and $u = b$ on the boundary $(\Gamma_\mathrm{D})_T \cap \mathfrak{A}_{\Gamma_\mathrm{D}}(F)$.*

A related result can be shown for the sequence $\{\mu_\delta\}$ by exploiting the estimates in Lemma 7.3.5.

Lemma 7.3.7 *Let a sequence $\{t_m\} \subseteq [0, T]$ containing T be dense. There exists a fine representation $\{U_k^m\}_{k \in \mathbb{N}}$ for $F(t_m)$ for every $m \in \mathbb{N}$, a function $\mu \in L_t^2 H_{x,\mathrm{loc}}^1(F)$ and a subsequence of $\{\mu_\delta\}$ (also denoted by $\{\mu_\delta\}$) such that for all $k, m \in \mathbb{N}$*

$$\mu_\delta \rightharpoonup \mu \quad \text{in } L^2(0, t_m; H^1(U_k^m)) \qquad (7.27)$$

as $\delta \to 0^+$.

Proof. A fine representation $\{U_k^m\}_{k \in \mathbb{N}}$ of $F(t_m)$ can be constructed by countably many open cubes $Q \subset\subset F(t_m) \cap \Omega$ and of finitely many open sets of the form $U \cap \Omega$ such that U satisfies (7.22) from Lemma 7.3.5 (ii). For each $k, m \in \mathbb{N}$, we have the estimate

$$\|\mu_\delta\|_{L^2(0,t;H^1(U_k^m))} \leq C$$

for all $0 < \delta \ll 1$ by Lemma 7.3.5. By successively choosing sub-sequences and by a diagonalizing argument, we obtain a $\mu \in L_t^2 H_{x,\mathrm{loc}}^1(F)$ such that (7.27) is satisfied (cf.

proof of Lemma 6.3.13). $\square$

The a-priori estimates and the convergence properties of $\{z_\delta\}$ in Lemma 7.3.4 and of $\{\mu_\delta\}$ in Lemma 7.3.7, respectively, yield the following corollary.

Corollary 7.3.8 *It holds for $\delta \to 0^+$:*

$$\mathrm{M}(z_\delta)\nabla\mu_\delta \rightharpoonup \mathrm{M}(z)\nabla\mu \text{ in } L^2(F;\mathbb{R}^n),$$
$$\mathrm{M}(z_\delta)\nabla\mu_\delta \to 0 \text{ in } L^2(\Omega_T \setminus F;\mathbb{R}^n).$$

Now, we have all the necessary convergence properties to perform the degenerate limit in (7.15)-(7.20).

The degenerate limit $\delta \to 0^+$ can be performed as follows:

- We define the strain by $e := \widehat{e}|_F \in L^2(F;\mathbb{R}^{n\times n})$ and obtain for the remaining variables

$$c \in L^\infty(0,T;H^1(\Omega)) \cap H^1(0,T;(H^1(\Omega))^*), \quad u \in L^2_t H^1_{x,\mathrm{loc}}(\mathfrak{A}_{\Gamma_\mathrm{D}}(F);\mathbb{R}^n),$$
$$z \in L^\infty(0,T;W^{1,p}(\Omega)) \cap H^1(0,T;L^2(\Omega)), \quad \mu \in L^2_t H^1_{x,\mathrm{loc}}(F)$$

  with $e = \epsilon(u)$ in $\mathfrak{A}(F)$.

- Passing to the limit $\delta \to 0^+$ in (7.15), (7.18) and (7.20) imply properties (7.6), (7.9) and (7.11) as in Chapter 6.

- Using Lemma 7.3.4 (a) and Corollary 7.3.8, we can pass to $\delta \to 0^+$ in (7.16) and obtain (7.7).

  Let $\zeta \in L^2(0,T;H^1(\Omega))$ with $\mathrm{supp}(\zeta) \subseteq F$ be a test function. Furthermore, let $\{\psi_l\}$ be a partition of unity of the compact set $K := \mathrm{supp}(\zeta)$ according to Lemma 2.4.5. For each $l \in \mathbb{N}$, we obtain $\mathrm{supp}(\zeta\psi_l) \subseteq \overline{U_l^{m_l}} \times [0,t_{m_l}]$. Then, integrating (7.17) in time from 0 to t_{m_l}, testing the result with $\zeta\psi_l$ and passing to $\delta \to 0^+$ by using Lemma 7.3.4 and Lemma 7.3.7 show

$$\int_0^{t_m} \int_\Omega \mu\zeta\psi_l \,\mathrm{d}x\,\mathrm{d}s = \int_0^{t_m} \int_\Omega \left(\nabla c \cdot \nabla(\zeta\psi_l) + \partial_c W^{\mathrm{ch}}(c)\zeta\psi_l + \partial_c W^{\mathrm{el}}(c,\widehat{e},z)\zeta\psi_l\right) \mathrm{d}x\,\mathrm{d}s.$$

  Summing with respect to $l \in I$ and noticing $\sum_{l\in I}\psi_l \equiv 1$ on $\mathrm{supp}(\zeta)$ yield (7.8).

In conclusion, the limit procedure in this section yields functions (c,e,u,z,μ) with $e = \epsilon(u)$ in $\mathfrak{A}_{\Gamma_\mathrm{D}}(F)$ and which satisfy properties (ii)-(vi) of Definition 7.2.1. In particular, the damage function z has no jumps with respect to time. We cannot ensure that $\{z > 0\}$ equals $\mathfrak{A}_{\Gamma_\mathrm{D}}(\{z > 0\})$ and, moreover, if $F \setminus \mathfrak{A}_{\Gamma_\mathrm{D}}(\{z > 0\}) \neq \emptyset$, it is not clear whether u can be extended to a function on F such that $e = \epsilon(u)$ also holds in F. This issue is addressed in the next section where such limit functions are concatenated in order to obtain global-in-time approximate weak solutions by Zorn's lemma.

7.3.3   Local- and global-in-time existence results

In this subsection, we are going to prove the main results of this chapter

Proof of Theorem 7.2.5. Zorn's lemma can be applied to the set

$$\mathcal{P} := \big\{ (\widehat{T}, c, u, z, \mu) \,|\, 0 < \widehat{T} \leq T \text{ and } (c, u, z, \mu) \text{ is a weak solution on}$$
$$[0, \widehat{T}] \text{ according to Definition 7.2.1} \big\}$$

to find a maximal element with respect to the following partial ordering

$$(\widehat{T}_1, c_1, u_1, z_1, \mu_1) \leq (\widehat{T}_2, c_2, u_2, z_2, \mu_2) \quad \Leftrightarrow \quad \widehat{T}_1 \leq \widehat{T}_2, \; c_2|_{[0,\widehat{T}_1]} = c_1, \; u_2|_{[0,\widehat{T}_1]} = u_1,$$
$$z_2|_{[0,\widehat{T}_1]} = z_1, \; \mu_2|_{[0,\widehat{T}_1]} = \mu_1. \qquad (7.28)$$

Indeed, $\mathcal{P} \neq \emptyset$ by the result in Section 7.3.2. More precisely, since $z \in L^\infty(0, T; W^{1,p}(\Omega)) \cap H^1(0, T; L^2(\Omega))$ and since $0 < \kappa \leq z^0$, we find an $\delta > 0$ such that $\{z(t) > 0\} = \mathfrak{A}_{\Gamma_D}(\{z(t) > 0\})$ for all $t \in [0, \delta]$. For the proof that every totally ordered subset of $\mathcal{P}$ has an upper bound, we refer to Lemma 6.3.15. $\qquad \square$

Lemma 7.3.9 *Let $t_1 < t_2 < t_3$ be real numbers and let $\eta > 0$. Suppose that*

$\widetilde{q} := (\widetilde{c}, \widetilde{e}, \widetilde{u}, \widetilde{z}, \widetilde{\mu}, \widetilde{F})$ *is an approximate weak solution on* $[t_1, t_2]$,

$\widehat{q} := (\widehat{c}, \widehat{e}, \widehat{u}, \widehat{z}, \widehat{\mu}, \widehat{F})$ *is an approximate weak solution on* $[t_2, t_3]$

     with $\widehat{\mathfrak{e}}_{t_2}^+ = \mathfrak{E}(\widehat{c}(t_2), \widehat{b}(t_2), \widehat{z}^+(t_2))$ (the value $\mathfrak{e}_{t_2}^+$ for $\widehat{q}$ in Definition 7.2.1).

Furthermore, suppose the compatibility condition

$$\widehat{c}(t_2) = \widetilde{c}(t_2),$$
$$\widehat{z}^+(t_2) = \widetilde{z}^-(t_2) \mathbb{1}_{\mathfrak{A}_{\Gamma_D}(\{\widetilde{z}^-(t_2)>0\})}$$

and the Dirichlet boundary data $b \in W^{1,1}(t_1, t_3; W^{1,\infty}(\Omega; \mathbb{R}^n))$.

     Then, we obtain that $q := (c, e, u, z, \mu, F)$ defined as $q|_{[t_1,t_2)} := \widetilde{q}$ and $q|_{[t_2,t_3]} := \widehat{q}$ is an approximate weak solution on $[t_1, t_3]$.

Proof. Because of the properties in Remark 7.3.2 we can prove the following crucial energy estimate at time point t_2:

$$\lim_{s \to t_2^-} \operatorname*{ess\,inf}_{\tau \in (s, t_2)} \mathcal{E}(c(\tau), e(\tau), z(\tau)) = \lim_{s \to t_2^-} \operatorname*{ess\,inf}_{\tau \in (s, t_2)} \mathcal{E}(c(\tau), e(\tau), z^-(\tau))$$

$$\geq \lim_{s \to t_2^-} \operatorname*{ess\,inf}_{\tau \in (s, t_2)} \mathcal{E}(c(\tau), \epsilon(u(\tau)), z^-(\tau) \mathbb{1}_{\mathfrak{A}_{\Gamma_D}(\{z^-(\tau)>0\})})$$

$$\geq \lim_{s \to t_2^-} \operatorname*{ess\,inf}_{\tau \in (s, t_2)} \mathfrak{E}(c(\tau), b(\tau), z^-(\tau) \mathbb{1}_{\mathfrak{A}_{\Gamma_D}(\{z^-(\tau)>0\})})$$

$$\geq \mathfrak{E}(c(t_2), b(t_2), \chi)$$

$$\geq \mathfrak{E}(c(t_2), b(t_2), z^+(t_2))$$

with $\chi := z^-(t_2)\mathbb{1}_{\bigcap_{\tau \in (t_1, t_2)} \mathfrak{A}\Gamma_D(\{z^-(\tau)>0\})}$. For all details, we refer to Lemma 6.3.16. $\quad\square$

Proof of Theorem 7.2.4. This result can also be proven by using Zorn's lemma on the set

$$\mathcal{P} := \big\{(\widehat{T}, c, e, u, z, \mu, F) \,|\, 0 < \widehat{T} \leq T \text{ and } (c, e, u, z, \mu, F) \text{ is an approximate weak}$$
$$\text{solution on } [0, \widehat{T}] \text{ with fineness } \eta \text{ according to Definition 7.2.3}\big\}$$

with an ordering analogously to (7.28). The assumptions for Zorn's lemma can be proven as in Theorem 7.2.5 (see proof of Theorem 6.2.6). To show that a maximal element from $\mathcal{P}$ is actually an approximate weak solution on the time-interval $[0, T]$, we need the concatenation property in Lemma 7.3.9. Indeed, if a maximal element $\widetilde{q}$ is only defined on an time-interval $[0, \widetilde{T}]$ with $\widetilde{T} < T$ we can apply the degenerated limit procedure in Section 7.3.2 to the initial values $c(\widetilde{T})$ and $z(\widetilde{T})$ to obtain a new limit function $\widehat{q}$. By exploiting Lemma 7.3.9, q is an approximate weak solution on the time-interval $[0, \widetilde{T}+\delta]$ for a small $\delta > 0$ which contradicts the maximality of $\widetilde{q}$. $\quad\square$

In this work, we have investigated mathematical models describing both phenomena, phase separation and damage processes, in a unifying approach. Phase separation is modeled by elastic Cahn-Hilliard equations, whereas the damage processes are described by a doubly nonlinear differential inclusion. The forces are assumed to be in a quasi-static equilibrium. We have introduced the corresponding PDE model (1.1)-(1.2) in Chapter 3 and we have shown thermodynamic consistency. The main aim has been to prove existence of weak solutions of the coupled system for various types of free energy densities of the form (1.3). The damage dissipation potential is rate-dependent and specified by (1.4).

The following cases depending on the choice of the free energy have been studied:

(i) At first, two-component Cahn-Hilliard systems coupled with incomplete damage processes have been investigated. The associated free energy functional reads as

$$\int_\Omega \left(\frac{1}{2}|\nabla c|^2 + \frac{1}{p}|\nabla z|^p + W^{\text{ch}}(c) + W^{\text{el}}(c, \epsilon(u), z) + I_{[0,\infty)}(z) \right) dx,$$

with $p > n$ (n is the space dimension). Here, W^{ch} is the chemical energy density fulfilling a polynomial growth condition and W^{el} the elastic energy density which covers homogeneous elasticity of the form

$$W^{\text{el}}(c, e, z) = \frac{1}{2}(g(z) + \delta)(e - e^\star(c)) : \mathbf{C}(e - e^\star(c))$$

with $\delta > 0$. It is also possible to incorporate a damage dependent potential f in the free energy. The mobility tensor in the diffusion equation (1.1a) is assumed to be constant.

In Chapter 4, a notion of weak solutions consisting of a variational inequality and an energy inequality have been developed (see Definition 4.2.6) and global-in-time existence have been proven in Theorem 4.2.7. To this end, we have studied a time-discrete version of the coupled system. An approximation and a variational technique have been established (see Lemma 2.3.18 and Lemma 2.3.19) to handle the passage from the time-discrete to the time-continuous regime.

(ii) In the next step, the previous existence results have been extended to multi-component Cahn-Hilliard systems coupled with damage processes where the free energy is specified by

$$\int_\Omega \left(\frac{1}{2}\Gamma\nabla c : \nabla c + \frac{1}{2}|\nabla z|^2 + W^{\text{ch}}(c) + W^{\text{el}}(c, \epsilon(u), z) + I_{[0,\infty)}(z) \right) \mathrm{d}x.$$

The growth assumptions are now less restrictive (see (5.2)). In particular, we allow inhomogeneous elastic energy densities of the type

$$W^{\text{el}}(c, e, z) = \frac{1}{2}(g(z) + \delta)(e - e^\star(c)) : \mathbf{C}(c)(e - e^\star(c))$$

with $\delta > 0$. The chemical potential can now be of polynomial or logarithmic type (see (5.3)).

We treated this case in Chapter 5, where we slightly weakened the notion of weak solutions in comparison to the notion in the case (i) (see Definition 5.2.3). Beyond that, a higher integrability result for the strain tensor has been established to deal with logarithmic chemical potentials (see Theorem 5.3.10). We proved existence of global-in-time weak solutions in Theorem 5.2.6 and in Theorem 5.2.7.

(iii) Based on the previous results, we have turned our attention to complete damage processes with a quasi-static force balance law. In the mathematical literature, global-in-time existence have been studied either for rate-independent [BMR09] or for viscoelastic systems [MRZ10, RR12] so far. The free energy we have considered reads as

$$\int_\Omega \left(\frac{1}{p}|\nabla z|^p + W^{\text{el}}(\epsilon(u), z) + I_{[0,\infty)}(z) \right) \mathrm{d}x$$

with $p > n$. The elastic energy density W^{el} has the structure

$$W^{\text{el}}(e, z) = \frac{1}{2}g(z)e : \mathbf{C}e.$$

New ideas have been developed to handle this degenerating elastic energy. For instance, the PDE system is only imposed on that material parts which are connected to the Dirichlet boundary (see Definition 3.3.1). We developed a notion of weak solutions (see Definition 6.2.1 and Theorem 6.2.3) and approximate weak solutions (see Definition 6.2.4) within suitable trajectory spaces. Finally, we proved maximal local-in-time existence of weak solutions in Theorem 6.2.7 as well as global-in-time existence of approximate weak solutions in Theorem 6.2.6.

(iv) In the last case, we have brought together the ideas from the previous approaches in (i) and (iii). More specifically, elastic Cahn-Hilliard equations coupled with complete damage processes are considered (see Definition 3.4.3). A further novelty is that the diffusion mobility tensor now depends on the damage variable and is degenerating when the damage process is complete. The free energy is given by

$$\int_\Omega \left(\frac{1}{2}|\nabla c|^2 + \frac{1}{p}|\nabla z|^p + W^{\mathrm{ch}}(c) + W^{\mathrm{el}}(c, \epsilon(u), z) + f(z) + I_{[0,\infty)}(z) \right) \mathrm{d}x$$

with $p > n$. The elastic energy density W^{el} covers homogeneous cases as in (i), i.e.,

$$W^{\mathrm{el}}(c, e, z) = \frac{1}{2}g(z)(e - e^\star(c)) : \mathbf{C}(e - e^\star(c)),$$

but is degenerating when the damage is complete (here, $\delta = 0$). The chemical energy density W^{ch} fulfills a polynomial growth condition as in (i).

Employing the same approach as in (iii), we proved maximal local-in-time existence of weak solutions in Theorem 7.2.5 as well as global-in-time existence of approximate weak solutions in Theorem 7.2.4.

To our best knowledge, the PDE systems considered in (i)-(iv) have not been considered in the mathematical literature so far.

In the end, we would like to mention that the results in this thesis might be extended to various directions in future works. We give some examples.

- Combining the ideas in (ii) with (iii), one could study multi-component Cahn-Hilliard equations with logarithmic chemical potentials coupled with complete damage processes. An additional difficulty would be the inhomogeneous elasticity as considered in (ii) because then e enters the term $\partial_c W^{\mathrm{el}}$ in the Cahn-Hilliard equation quadratically. A notion of weak solutions involving *generalized Young measures* might be employed (see [DM87]).

- A further interesting case would be damage processes in viscoelastic media coupled with phase separation. Especially in the complete damage regime, better results are expected because the L^2-norm of the displacement field and the L^2-norm of the strain field can be locally controlled. Therefore, a Korn type inequality can be applied on smooth subsets. Eventually, we could even describe the not completely damaged material parts which are disconnected from the Dirichlet boundary and the concept of approximate weak solutions would reduce to weak solutions. Damage processes in thermoviscoelastic solids are already studied in [RR12].

- Another interesting consideration would be to consider systems which couple phase separation with rate-independent damage processes, i.e., with a dissipation potential (1.4) with $\beta = 0$. One might think about the usage of the notion of *energetic solutions* (within the framework of rate-independent systems; see [Mie05]) for the differential inclusion (1.1c) to this case.

References

[AB07] C.D. Aliprantis and K.C. Border. *Infinite Dimensional Analysis: A Hitchhiker's Guide.* Springer, 2007.

[AC79] S.M. Allen and J.W. Cahn. A microscopic theory for antiphase boundary motion and its application to antiphase domain coarsening. *Acta Metal.*, 27:1085–1095, 1979.

[AFP00] L. Ambrosio, N. Fusco, and D. Pallara. *Functions of Bounded Variation and Free Discontinuity Problems.* Oxford University Press Inc., New York, 2000.

[Alt99] H.W. Alt. *Lineare Funktionalanalysis.* Springer-Verlag Heidelberg, 1999.

[AP92] H.W. Alt and I. Pawłow. Existence of solutions for non-isothermal phase separation. *Adv. Math. Sci. Appl.*, 1(2):319–409, 1992.

[Bab11] J.-F. Babadjian. A quasistatic evolution model for the interaction between fracture and damage. *Arch. Ration. Mech. Anal.*, 200(3):945–1002, 2011.

[BB99] J.W. Barrett and J.F. Blowey. Finite element approximation of the Cahn-Hilliard equation with concentration dependent mobility. *Math. of Computation*, 68(226):487–517, 1999.

[BB11] A. Berti and I. Bochicchio. A mathematical model for phase separation: a generalized Cahn-Hilliard equation. *Math. Methods Appl. Sci.*, 34(10):1193–1201, 2011.

[BBR12] E. Bonetti, G. Bonfanti, and R. Rossi. Analysis of a unilateral contact problem taking into account adhesion and friction. *J. Differ. Equations*, 253(2):438–462, 2012.

[BCD+02] E. Bonetti, P. Colli, W. Dreyer, G. Gilardi, G. Schimperna, and J. Sprekels. On a model for phase separation in binary alloys driven by mechanical effects. *Physica D*, 165:48–65, 2002.

[BdS04] J.L. Boldrini and P.N. da Silva. A generalized solution to a Cahn-Hilliard/Allen-Cahn system. *Electron. J. Differ. Equ. (electronic only)*, 1(126):1–24, 2004.

[Ber11] J.-M.E. Bernard. Density results in Sobolev spaces whose elements vanish on a part of the boundary. *Chin. Ann. Math., Ser. B*, 32(6):823–846, 2011.

[BGN07] J.W. Barrett, H. Garcke, and R. Nürnberg. A phase field model for the electromigration of intergranular voids. *Interfaces Free Bound.*, 9(2):171–210, 2007.

[BK98] S. M. Buckley and P. Koskela. New Poincaré inequalities from old. *Annales Academiæ Scientiarum Fennicæ Mathematica*, 23:251–260, 1998.

[BM10] S. Bartels and R. Müller. A posteriori error controlled local resolution of evolving interfaces for generalized Cahn–Hilliard equations. *Interfaces and Free Boundaries*, 12(1):45–73, 2010.

[BMR09] G. Bouchitte, A. Mielke, and T. Roubíček. A complete-damage problem at small strains. *ZAMP Z. Angew. Math. Phys.*, 60:205–236, 2009.

[BP05] L. Bartkowiak and I. Pawłow. The Cahn-Hilliard-Gurtin system coupled with elasticity. *Control and Cybernetics*, 34:1005–1043, 2005.

[Bra02] A. Braides. *Gamma-convergence for beginners*, volume 1. Oxford Lecture Series in Mathematics and its Applications 22. Oxford, 2002.

[Bra06] A. Braides. A handbook of Γ-convergence. volume 3 of *Handbook of Differential Equations: Stationary Partial Differential Equations*, pages 101 – 213. North-Holland, 2006.

[BS04] E. Bonetti and G. Schimperna. Local existence for Frémond's model of damage in elastic materials. *Contin. Mech. Thermodyn.*, 16(4):319–335, 2004.

[BSS05] E. Bonetti, G. Schimperna, and A. Segatti. On a doubly nonlinear model for the evolution of damaging in viscoelastic materials. *J. of Diff. Equations*, 218(1):91–116, 2005.

[BW05] T. Blesgen and U. Weikard. Multi-component Allen-Cahn equation for elastically stressed solid. *Electronic Journal of Differential Equations*, Paper no. 89, 2005.

[Cah61] J.W. Cahn. On spinodal decomposition. *Acta Metal.*, 9:795–801, 1961.

[Car86] A. Carpinteri. *Mechanical damage and crack growth in concrete. Plastic collapse to brittle fracture.* Springer, Netherlands, 1986.

[CGPGS10] P. Colli, G. Gilardi, P. Podio-Guidugli, and J. Sprekels. Existence and uniqueness of a global-in-time solution to a phase segregation problem of the Allen-Cahn type. *Math. Models Methods Appl. Sci.*, 20(4):519–541, 2010.

[CGPGS12] P. Colli, G. Gilardi, P. Podio-Guidugli, and J. Sprekels. Global existence for a strongly coupled Cahn–Hilliard system with viscosity. *Boll. Unione Mat. Ital.*, 5(9):495–513, 2012.

[CH58] J.W. Cahn and J.E. Hiliard. Free energy of a uniform system. i. interfacial free energy. *Journal of Chemical Physics*, 28:258–267, 1958.

[CMP00] M. Carrive, A. Miranville, and A. Piétrus. The Cahn-Hilliard equation for deformable elastic continua. *Adv. Math. Sci. Appl.*, 10(2):539–569, 2000.

[CP08] P. Cherfils and M. Pierre. Non-global existence for an Allen-Cahn-Gurtin equation with logarithmic free energy. *J. Evol. Equ.*, 8(4):727–748, 2008.

[CV02] D. Candeloro and A. Volčič. Radon-Nikodým Theorems. In E. Pap, editor, *Handbook of Measure Theory*, pages 249–294, Amsterdam, 2002. Elsevier.

[CVC12] S. Carillo, V. Valente, and G.V. Caffarelli. An existence theorem for the magneto-viscoelastic problem. *Discrete Contin. Dyn. Syst., Ser. S*, 5(3):435–447, 2012.

[DA91] A. R. Denton and N. W. Ashcroft. Vegard's law. *Phys. Rev. A*, 43:3161–3164, 1991.

[Din66] N. Dinculeanu. *Vector Measures*. VEB Deutscher Verlag der Wissenschaften, Berlin, GDR, 1966.

[Din02] N. Dinculeanu. Vector integration in Banach spaces and application to stochastic integration. In E. Pap, editor, *Handbook of Measure Theory*, pages 345–399, Amsterdam, 2002. Elsevier.

[DM87] R.J. DiPerna and A.J. Majda. Oscillations and Concentrations in Weak Solutions of the Incompressible Fluid Equations. *Commun. Math. Phys.*, 108:667–689, 1987.

[DM00] W. Dreyer and W.H. Müller. A study of the coarsening in tinlead solders. *Internat. J. Solids and Structures*, 37(28):3841–3871, 2000.

[DM01] W. Dreyer and W.H. Müller. Modeling diffusional coarsening in eutectic tinlead solders: A quantitative approach. *Internat. J. Solids and Structures*, 38(8):1433–1458, 2001.

[Dob07] M. Dobrowolski. *Angewandte Funktionalanalysis: Funktionalanalysis, Sobolev-Räume und Elliptische Differentialgleichungen.* Springer London, 2007.

[EL91] C.M. Elliott and S. Luckhaus. A generalised diffusion equation for phase separation of a multi-component mixture with interfacial free energy. *IMA Preprint no. 195*, 1991.

[Ell89] C.M. Elliott. The Cahn-Hilliard model for the kinetics of phase separation. *International Series of Numerical Mathematics*, 37(28):3841–3871, 1989.

[EM06] M. A. Efendiev and A. Mielke. On the rate-independent limit of systems with dry friction and small viscosity. *J. Convex Analysis*, 13:151–167, 2006.

[FG06] G.A. Francfort and A. Garroni. A variational view of partial brittle damage evolution. *Arch. Ration. Mech. Anal.*, 182(1):125–152, 2006.

[FK09] J. R. Fernández and K. L. Kuttler. An existence and uniqueness result for an elasto-piezoelectric problem with damage. *Math. Mod. Meth. Appl. Sci.*, 19(1):31–50, 2009.

[FN96] M. Frémond and B. Nedjar. Damage, gradient of damage and principle of virtual power. *Int. J. Solids Structures*, 33(8):1083–1103, 1996.

[Fré02] M. Frémond. *Non-smooth thermomechanics.* Berlin: Springer, 2002.

[Fré12] M. Frémond. *Phase Change in Mechanics.* Lecture Notes of the Unione Matematica Italiana. Springer, 2012.

[Gar00] H. Garcke. *On mathematical models for phase separation in elastically stressed solids.* Habilitation thesis, University Bonn, 2000.

[Gar05a] H. Garcke. Mechanical Effects in the Cahn-Hilliard Model: A Review on Mathematical Results. In A. Miranville, editor, *Mathematical Methods and Models in phase transitions*, pages 43–77. Nova Science Publ., 2005.

[Gar05b] H. Garcke. On a Cahn-Hilliard model for phase separation with elastic misfit. *Annales de l'Institut Henri Poincaré (C) Non Linear Analysis*, 22(2):165 – 185, 2005.

[Gia83] M. Giaquinta. *Multiple integrals in the calcula of variations and nonlinear elliptic systems.* Annals of Mathematical Studies. Princeton University Press, 1983.

[Gia05] A. Giacomini. Ambrosio–Tortorelli approximation of quasi–static evolution of brittle fractures. *Calc. Var. Partial Differ. Equ.*, 22(2):129–172, 2005.

[GL09] A. Garroni and C. Larsen. Threshold-based quasi-static brittle damage evolution. *Arch. Ration. Mech. Anal.*, 194(2):585–609, 2009.

[GR07] G. Gilardi and E. Rocca. Well-posedness and long-time behaviour for a singular phase field system of conserved type. *IMA journal of applied mathematics*, 72(4):498 – 530, 2007.

[GRW01] H. Garcke, M. Rumpf, and U. Weikard. The Cahn-Hilliard equation with elasticity: Finite element approximation and qualitative studies. *Interfaces Free Bound.*, 3:101–118, 2001.

[GUE+07] M.G.D. Geers, R.L.J.M. Ubachs, M. Erinc, M.A. Matin, P.J.G. Schreurs, and W.P. Vellinga. Multiscale Analysis of Microstructura Evolution and Degradation in Solder Alloys. *Internatilnal Journal for Multiscale Computational Engineering*, 5(2):93–103, 2007.

[Gur96] M.E. Gurtin. Generalized Ginzburg-Landau and Cahn-Hilliard equations based on a microforce balance. *Physica D*, 92:178–192, 1996.

[GW05] H. Garcke and U. Weikard. Numerical approximation of the Cahn-Larché equation. *Numer. Math.*, 100(4):639–662, 2005.

[HCW91] P. G. Harris, K. S Chaggar, and M. A. Whitmore. The Effect of Ageing on the Microstructure of 60:40 Tin–lead Solders. *Soldering & Surface Mount Technology Improved physical understanding of intermittent failure in continuous*, 3:20–33, 1991.

[HK11] C. Heinemann and C. Kraus. Existence of weak solutions for Cahn-Hilliard systems coupled with elasticity and damage. *Adv. Math. Sci. Appl.*, 21(2):321–359, 2011.

[HK12a] C. Heinemann and C. Kraus. Complete damage in linear elastic materials — modeling, weak formulation and existence results. *WIAS preprint no. 1722 (submitted)*, 2012.

[HK12b] C. Heinemann and C. Kraus. A degenerating Cahn-Hilliard system coupled with complete damage processes. *WIAS preprint no. 1759*, 2012.

[HK13] C. Heinemann and C. Kraus. Existence results for diffuse interface models describing phase separation and damage. *Eur. J. Appl. Math.*, 24(2):179–211, 2013.

[HR99] W. Han and B.D. Reddy. *Plasticity: Mathematical Theory and Numerical Analysis*, volume 9 of *Interdisciplinary Applied Mathematics*. Springer-Verlag, 1999.

[KN94] N. Kenmochi and M. Niezgódka. Nonlinear system for non-isothermal diffusive phase separation. *J. Math. Anal. Appl.*, 188(2):651–679, 1994.

[KRZ11] D. Knees, R. Rossi, and C. Zanini. *A vanishing viscosity approach to a rate-independent damage model*. WIAS preprint no. 1633. WIAS, 2011.

[LC82] F.C. Larché and J.W. Cahn. The effect of self-stress on diffusion in solids. *Acta Metal.*, 30:1835–1845, 1982.

[LD05] J. Lemaitre and R. Desmorat. *Engineering Damage Mechanics: Ductile, Creep, Fatigue and Brittle Failures*. Springer-Verlag, Berlin, 2005.

[LSC+04] J. Lau, D. Shangguan, T. Castello, R. Horsley, J. Smetana, N. Hoo, W. Dauksher, D. Love, I. Menis, and B. Sullivan. Failure analysis of lead-free solder joints for high-density packages. *Soldering & Surface Mount Technology*, 16(2):69–76, 2004.

[Mas93] G.D. Maso. *Introduction to Gamma-Convergence*. Progress in Nonlinear Differential Equations and Their Applications. Birkhäuser, 1993.

[Mer05] T. Merkle. The Cahn-Larché system: A model for spinodal decomposition in eutectic solder; modelling, analysis and simulation. PhD-thesis, Universität Stuttgart, Stuttgart, 2005.

[Mie95] C. Miehe. Discontinuous and continuous damage evolution in Ogden-type large-strain elastic materials. *Eur. J. Mech.*, 14:697–720, 1995.

[Mie05] A. Mielke. Evolution in rate-independent systems. *Handbook of Differential Equations: Evolutionary Equations*, 2:461–559, 2005.

[Mie11] A. Mielke. Complete-damage evolution based on energies and stresses. *Discrete Contin. Dyn. Syst., Ser. S*, 4(2):423–439, 2011.

[MK00] C. Miehe and J. Keck. Superimposed finite elastic-viscoelastic-plastoelastic stress response with damage in filled rubbery polymers. Experiments, modelling and algorithmic implementation. *J. Mech. Phys. Solids*, 48:323–365, 2000.

[MR06] A. Mielke and T. Roubíček. Rate-independent damage processes in nonlinear elasticity. *Mathematical Models and Methods in Applied Sciences*, 16:177–209, 2006.

[MRS12] A. Mielke, R. Rossi, and G. Savaré. BV solutions and viscosity approximations of rate-independent systems. *ESAIM: Control, Optimisation and Calculus of Variations*, 18:36–80, 2012.

[MRZ10] A. Mielke, T. Roubíček, and J. Zeman. Complete Damage in elastic and viscoelastic media. *Comput. Methods Appl. Mech. Engrg*, 199:1242–1253, 2010.

[MS05] A. Miranville and G. Schimperna. Nonisothermal phase separation based on a microforce balance. *Discrete Contin. Dyn. Syst., Ser. B*, 5(3):753–768, 2005.

[MS11] A. Menzel and P. Steinmann. A theoretical and computational framework for anisotropic continuum damage mechanics at large strains. *Int. J. Solids Struct.*, 38:9505–9523, 2011.

[MT99] A. Mielke and F. Theil. A mathematical model for rate-independent phase transformations with hysteresis. In R. Balean H.-D. Alber and R. Farwig, editors, *Models of Continuum Mechanics in Analysis and Engineering*, pages 117–129, Aachen, 1999. Shaker Verlag.

[MT10] A. Mielke and M. Thomas. Damage of nonlinearly elastic materials at small strain — Existence and regularity results. *ZAMM Z. Angew. Math. Mech*, 90:88–112, 2010.

[MV87] J.J. Moreau and M. Valadier. A chain rule involving vector functions of bounded variation. *Journal of Functional Analysis*, 74(2):333–345, 1987.

[Nef02] P. Neff. On Korn's first inequality with non-constant coefficients. *Proc. R. Soc. Edinb., Sect. A, Math.*, 132(1):221–243, 2002.

[Nir59] L. Nirenberg. On elliptic differential equations. *Ann. Scuola Norm. Pisa (III)*, 13:1–48, 1959.

[RR08] E. Rocca and R. Rossi. Analysis of a nonlinear degenerating PDE system for phase transitions in thermoviscoelastic materials. *J. Differ. Equations*, 245(11):3327–3375, 2008.

[RR12] E. Rocca and R. Rossi. A degenerating PDE system for phase transitions and damage. *arXiv:1205.3578v1*, 2012.

[Sim86] J. Simon. Compact sets in the space $L^p(0, T; B)$. *Annali di Matematica Pura ed Applicata*, 146:65–96, 1986.

[Sob38] S.L. Sobolev. On a theorem of functional analysis. *Mat. Sbornik*, 46:471–497, 1938.

[USG07] R.L.J.J. Ubachs, P.J.G. Schreurs, and M.G.D. Geers. Elasto-viscoplastic nonlocal damage modelling of thermal fatigue in anisotropic lead-free solder. *Mechanics of Materials*, 39:685–701, 2007.

[Vra03] I.I. Vrabie. *Co-Semigroups and Applications*. Elsevier Science, 2003.

[WBB06] Y. Wang, J. Bergström, and C. Burman. Four-point bending fatigue behaviour of an iron-based laser sintered material. *International Journal of Fatigue*, 28(12):1705–1715, 2006.

[Wei01] U. Weikard. Numerische Lösungen der Cahn-Hilliard-Gleichung und der Cahn-Larché-Gleichung. PhD-thesis, Universität Bonn, Bonn, 2001.

[Zei90] E. Zeidler. *Nonlinear Functional Analysis and Its Applications: II/A: Linear Monotone Operators*. Nonlinear Functional Analysis and Its Applications. Springer-Verlag, 1990.

[Zie89] W.P. Ziemer. *Weakly differentiable functions*. Springer-Verlag New York, Inc., 1989.

List of Figures

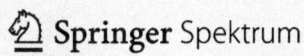